EVOKED POTENTIAL MANUAL

EVOKED POTENTIAL MANUAL

a practical guide to clinical applications

E. Colon

St. Radboud Hospital, University of Nijmegen
The Netherlands

S. Visser

Free University Hospital, Amsterdam
The Netherlands

J. de Weerd

St. Radboud Hospital, University of Nijmegen, The Netherlands, and
University of Wisconsin, Madison WI, USA

A. Zonneveldt

St. Lucas Hospital, Amsterdam
The Netherlands

1983 Springer-Science+Business Media, B.V.

Library of Congress Cataloging in Publication Data

Main entry under title:

Evoked potential manual.

Includes bibliographical references and index.
1. Evoked potentials (Electrophysiology) 2. Nervous system--Diseases--Diagnosis. I. Colon, E.
RC386.6.E86E96 1983 616.8'047547 83-19524

ISBN 978-94-017-1501-0 ISBN 978-94-017-1499-0 (eBook)
DOI 10.1007/978-94-017-1499-0

PREFACE

K. Hecox.

A number of significant challenges confront the clinincian interested in pathology of the human nervous system. Because the many components of the nervous system are largely inaccessible to palpation, percussion, auscultation, or direct visual inspection, the traditional tools of clinical medicine are restricted in the amount of information they can provide. The neurologic examination certainly provides essential data on the integrity of the nervous system, especially with regard to motor function, but it is increasingly apparent that the quantitative analysis of disorders of the human nervous system requires greater sensitivity and specificity than is achieved in a routine neurologic examination. The major advances in imaging procedures (CT scanning, PET scanning, and nuclear magnetic resonance studies) are important new sources of information on neural integrity.

Important advances have also occurred in clinical neurophysiology. The development of single unit recordings in electromyography, the introduction of quantitative spectral analysis in EEG recordings, and the growth of sensory evoked potentials, are three examples of new techniques which provide previously unavailable information. The development of these techniques relies heavily on thechnical advances such as the increased sophistication in electronics, more powerful computer capabilities, and improvements in amplifier technologies. The widespread availability of microprocessors and electronic microcircuitry assures the continued growth and improved flexibility of recording and stimulating techniques in clinical neurophysiology. These advances have made new demands on neurodiagnosticians. The need for familiarity with electronics, computer sciences, and applied physics, is rapidly increasing. An understanding of the strengths and limitations of newer clinical procedures requires at least some familiarity with the scientific foundation upon which these procedures are based.

The need to train residents and re-educate faculty in these procedures and technologies is apparent. It is in this sense that this volume is particularly welcome in the clinical neurophysiology literature. Book outlines several clinical procedures providing important information on their limitations and strengths. The authors combine clinical and technical expertise, interfacing the medical and technological aspects of these procedures. The need for such a manual is apparent to all concerned with physician and technician training. Future training manuals must combine a sound understanding of the basic sciences underlying clinical neurophysiology with a broad experience in human neurologic disorders. This volume represents the successful confluence of these two forces, and as such sould represent a helpful addition to the instructional armamentarium of clinical neurophysiologists.

Kurt Hecox, M.D., Ph.D
Associate Professor
Dept. of Paediatrics and Neurology
University of Wisconsin Madison.

ACKNOWLEDGEMENTS

We are indebted to Ilse Vermeulen and Vincent van Ammers for preparing the camera-ready version of this book.

CONTENTS

CONTRIBUTORS

Blom, J.L., Dr.
Neth. Institute of Preventive Health Care, TNO,
Leiden,
The Netherlands.

Colon, E.J., Dr.
Department of clinical neurophysiology,
Institute of Neurology,
St. Radboud University Hospital,
Nijmegen,
The Netherlands.

van Lith, G.H.M., Dr.
Department of Electro-ophthalmology,
Eye Hospital,
Rotterdam,
The Netherlands.

Posthumus Meyjes, F.E., Dr.
Department of clinical neurophysiology,
Academic Medical Centre,
Amsterdam,
The Netherlands.

Rotteveel, J.J.
Department of Neuropediatrics,
Institute of Pediatrics,
St. Radboud University Hospital,
Nijmegen,
The Netherlands.

Thijssen, J.M., Dr. Ir.
Biophysical Laboratory,
St. Radboud University Hospital,
Nijmegen,
The Netherlands.

Visser, S.L., Prof. Dr.
Department of clinical neurophysiology,
Valerius Clinic / Free University,
Amsterdam,
The Netherlands.

de Weerd, J.P.C., Dr. Ir.
Department of clinical neurophysiology,
Institute of Neurology,
St. Radboud University Hospital,
Nijmegen,
The Netherlands.

Zonneveldt, A.
Department of clinical neurophysiology,
St. Lucas Hospital,
Amsterdam,
The Netherlands.

Chapter I

Introduction

J.P.C. de Weerd

Evoked potentials are the electrical voltage fluctuations which can be recorded from parts of the nervous system in response to stimulation of sensory modalities. One may distinguish between evoked potentials from the peripheral and the central nervous system. For the latter type a further subdivision can be made into spinal, brainstem, and cortical evoked potentials, according to the (assumed) structures from which the responses derive.

Another possible subdivision can be made with respect to the specific sensory modality which is stimulated. Accordingly, one has auditory, somatosensory, visual, gustatory and olfactory evoked potentials. At the present time, the former three types of evoked potentials are the ones that are commonly measured in diagnostic procedures. The corresponding sensory systems are relatively easy to stimulate, for example by means of an acoustic click, a brief electrical shock or a reversing light pattern. In contrast, stimulation of the olfactory and gustatory systems has proven to be technically and physiologically difficult and research in these areas is still in an early stage.

Historically, the study of human evoked potentials began shortly after Hans Berger recorded the first human electroencephalogram (Berger, 1929, 1930; Adrian and Matthews, 1934; Adrian and Yamagiwa, 1935). But it was not until 1947 that Dawson could unequivocally demonstrate that stimulus-specific, cortical evoked potentials could be recorded from the scalp (Dawson,1947).

Curiously enough, these early evoked potentials were recorded after stimulation of the somatosensory system, while later on major research efforts in evoked potentials mainly concentrated on the visual and, still later, on the auditory system.

The measurement of evoked potentials thus has a relatively short history. One likely explanation for this is the fact that cortical evoked potentials are usually small in amplitude as compared to the ongoing spontaneous EEG and cannot be visualized unless one employs some form of signal enhancement which had been unavailable for a long time.
Dawson's first evoked potential recording was made possible by means of the photographic superimposition technique, by which evoked potentials were enhanced through the superimposition of successive responses on a cathode ray oscilloscope. In subsequent years Dawson introduced the principles of signal averaging into neurophysiology and described an ingenious electro-mechanical machine for performing this operation automatically (Dawson, 1951; 1954). In the latter paper the capabilities of the method for extracting cortical sensory evoked potentials from an overwhelming background activity were clearly demonstrated. This major development was elaborated in a contribution by Barlow (1957), describing an electronic averager, based on an analog crosscorrelator system. Subsequently, Clark et al. (1961) described a digitally operated device (the Average Response Computer), and Clark and Molnar (1964) presented the Laboratory Instrument Computer, which can be regarded as the forerunner of the laboratory computers of today.

A breakthrough to the more widespread application of evoked potentials was formed by the commercial availability of the Computer of Average Transients and similar equipment produced in the early sixties. These technological advances initiated many studies on the characteristics, the origin and the usefulness of cortical evoked potentials in the fields of neurophysiology, clinical medicine, psychophysics and psychology.

Table 1.1 illustratively lists some of these applications, which will be dealt with more extensively in the subsequent chapters.

Against this background, the purpose of this book is to make the novice in the evoked potential field familiar with basic concepts, practical procedures and major clinical applications of auditory, somatosensory and visual evoked potentials. In addition the book is intended as a practical manual both for the neurologist and the technician involved in daily clinical evoked potential work. Therefore, the emphasis is on well-established concepts and clinical procedures, rather than on the state-of-the-art of the rapidly evolving and expanding evoked potential field.

Diagnosis and differentiation of disorders of the peripheral and central nervous system
- demyelinating diseases
- tumor localization in spinal cord and intracranial pathways
- inflammatory and traumatic lesions
- spinal and intracranial neuropathies

Evaluation and prognosis of mental function
- in congenital disorders
- in degenerative disorders

Estimation of hearing
- particularly in newborns and infants
- objective audiometry

Estimation of visual acuity
- particularly in infants
- objective optometry

Intra-operative monitoring
- during neurosurgical procedures
- during orthopedic procedures
- during open-heart surgery
- for controlling anaesthetic depth

Intensive Care monitoring
- post-operative
- in coma
- in assessing cerebral death

Table 1.1 Some typical applications of Evoked Potentials

Consequently, significant progress in the study of visual evoked potentials was made after the introduction of patterned light stimuli (Spehlmann, 1965; Spekreyse, 1966; Rietveld et al., 1967; Spekreyse et al., 1973). Unlike evoked potentials elicited by diffuse flashes, these "pattern evoked potentials" proved to be reliable both in form and latency between subjects (Harter and White, 1968) and to be useful in clinical practice (Halliday et al., 1972; Desmedt, 1977b).

A subsequent landmark in the history of evoked potentials was the discovery that short-latency brainstem components of the auditory evoked potential could be recorded from the human scalp (Sohmer and Feinmesser, 1967; Jewett et al., 1970; Jewett and Williston, 1971). The past decade has shown great progress in the understanding of the origin and the clinical usefulness of these early potentials.

Similar developments for somatosensory evoked potentials demonstrated the possibilities and usefulness of recording spinal and early brainstem components (Liberson and Kim, 1963; Cracco, 1972; 1973; Desmedt, 1980). More recently, short-latency visual evoked potentials have been described as well (van Hasselt, 1972; Cracco and Cracco, 1978; and others). As yet, their precise origin remains uncertain and their clinical significance is still to be established.

From the clinical point of view the measurement and analysis of evoked potentials has, during the past decade, reached a well established position within the arsenal of noninvasive electrodiagnostic procedures. Numerous papers in many fields of clinical medicine, notably in Clinical Neurophysiology, Neurology, and Neurosurgery, Audiology, Ophthalmology, Orthopedics, Pediatrics and Psychiatry, have described the clinical usefulness of evoked potentials in evaluating peripheral and central nervous system functioning, as well as auditory and visual functioning.

Excellent reviews covering the latter issue can be found in several recent books comprising selected papers and congress proceedings, such as Desmedt, 1977 a,b; Lehman and Callaway, 1979; Barber, 1980; Desmedt, 1980; Courjon et al., 1982; Stöhr et al., 1982 and Halliday, 1982.

REFERENCES

Adrian, E.D., Matthews, B.H.C.: The Berger rhythm: potential changes from the occipital lobes in man. Brain, 57: 355-385 (1934).

Adrian, E.D., Yamagiwa, K.: The origin of the Berger rhythm. Brain, 58: 323-351 (1935).

Barber, C. (ed.): Evoked Potentials, Proc. Int. Evoked Potentials Symp. Nottingham MTP Press. Falcon House Lancaster (1980).

Barlow, J.S.: An electronic method for detecting evoked responses of the brain and for reproducing their average wave forms. Electroenceph. Clin. Neurophysiol., 9: 340-343 (1957).

Berger, H.: Ueber das Elektrenkephalogramm des Menschen. Arch. Psychiat. Nervenkr. 87: 527-570 (1929). English translation in Gloor, P. (ed.): Hans Berger on the electroencephalogram of man. Electroenceph. Clin. Neurophysiol. Suppl. 28: 37-73 (1969).

Berger, H.: Ueber das Elektrenkephalogramm des Menschen. II. J. Physiol. Neurol. (Lpz.) 40: 160-179 (1930). English translation by P. Gloor: Hans Berger on the electroencephalogram of man. Electroenceph. Clin. Neurophysiol. Supp. 28: 75-93 (1969).

Clark, W.A., Brown, R.M., Goldstein, M.H., Molnar, C.E., O'Brien, D.F., Zieman, H.E.: The average response computer (ARC): a digital device for computing averages and amplitude and time histograms of electrophysiological response. IRE Trans. BME 8: 46-51 (1961).

Clark, W.A., Molnar, C.E.: The Linc: a description of the laboratory instrument computer. Ann. NY. Acad. Sci. 115: 653-668 (1964).

Courjon, J., Mauguiere, F., Revol, M. (eds.): Clinical applications of evoked potentials in neurology. Advances in Neurology, Vol. 32, Raven Press, New York (1982).

Cracco, R.Q.: The initial positive potential of the human scalp-recorded somatosensory evoked response. Electroenceph. clin. Neurophysiol. 32: 623-629 (1972).

Cracco, R.Q.: Spinal evoked response: peripheral nerve stimulation in man. Electroenceph. clin. Neurophysiol. 35: 379-386 (1973).

Cracco, R.Q., Cracco, J.B.: Visual evoked potential in man: early oscillatory potentials. Electroenceph. clin. Neurophysiol. 45: 731-739 (1978).

Dawson, G.D.: Cerebral responses to electrical stimulation of peripheral nerve in man. J. Neurol. Neurosurg. Psychiat. 10: 134-140 (1947).

Dawson, G.D.: A summation technique for detecting small signals in a large irregular background. J. Physiol. (Lond.), 115: 2P-3P (1951).

Dawson, G.D.: A summation technique for the detection of small evoked potentials. Electroenceph. clin. Neurophysiol. 6: 65-84 (1954).

Desmedt, J.E. (ed): Auditory evoked potentials in man. Psychopharmacology correlates of evoked potentials. Progress in Clin. Neurophysiol. vol. 2, Karger, Basel (1977a).

Desmedt, J.E. (ed.): Visual evoked potentials in man: new developments. Oxford University Press, Oxford (1977b).

Desmedt, J.E. (ed.): Clinical uses of cerebral, brainstem and spinal somatosensory evoked potentials. Progress in Clin. Neurophysiol. vol. 7, Karger, Basel (1980).

Halliday, A.M. (ed.): Evoked potentials in clinical testing. Churchill Livingstone, Edinburgh (1982).

Halliday, A.M., McDonald, W.I., Mushin, J.: Delayed visual evoked responses in optic neuritis. Lancet 1, 982-985 (1972).

Harter, M.R., White, C.T.: Effects of contour sharpness and check-size on visually evoked cortical potentials. Vision Res. 8: 701-711 (1968).

van Hasselt, P.: A short latency visual evoked potential recorded from the human mastoid process and auricle. Electroenceph. clin. Neurophysiol. 33: 517-519 (1972).

Jewett, D.L., Romano, M.N., Williston, J.S.: Human auditory evoked potentials: possible brainstem components detected on the scalp. Science 167: 1517-1518 (1970).

Jewett, D.L., Williston, J.S.: Auditory evoked far fields averaged from the scalp of humans. Brain 94: 681-696 (1971).

Lehmann, D., Callaway, E. (eds.): Human evoked potentials: applications and problems. Plenum Press, New York (1979).

Liberson, W.T., Kim, K.C.: The mapping out of evoked potentials elicited by stimulation of the median and peroneal nerves. Electroenceph. clin. Neurophysiol. 15: 721P (1963).

Rietveld, W.J., Tordoir, W.E., Hagenouw, J.R.B., Lubbers, J.A., Spoor, A.C.: Visual evoked responses to blank and to checkerboard patterned flashes. Acta Physiol. Pharmac. Neerl. 14: 259-285 (1967).

Sohmer, H., Feinmesser, M.: Cochlear action potentials recorded from the external ear in man. Ann. Otol. (St. Louis), 76: 427-435 (1967).

Spehlmann, R.: The averaged electrical responses to diffuse and to patterned light in the human. Electroenceph. clin. Neuropysiol. 19: 560-569 (1965).

Spekreyse, H.: Analysis of EEG responses to diffuse and to patterned light in the human. (Thesis) D.W. Junk, The Hague (1966).

Spekreyse, H., van der Tweel, L.H., Zuidema, T.: Contrast evoked responses in man. Vision Res. 13: 1577-1601 (1973).

Stöhr, M., Dickgans, J. Diener, H.C., Buettner, U.W.: Evozierte Potentialen. Springer, Berlin (1982).

Chapter II

The measurement of evoked potentials; technical aspects

J.P.C. de Weerd

1. INTRODUCTION AND SUMMARY OF THIS CHAPTER

When measuring evoked potentials we are faced with four distinct aspects:(i) the patient is subjected to some form of stimulation: auditory, somatosensory or visual; (ii) the electrical signals evoked by those stimuli are recorded and amplified; (iii) some form of signal processing is required because of the generally small responses relative to the background activity and, finally, (iv) the resulting wave forms are displayed, analysed and usually stored in some form for documentation and/or analysis.

Nowadays equipment in which all of the above functions have been integrated is commercially available. This fact, however, does not eliminate the need to have some basic understanding of the technical requirements and operating principles of evoked potential machines. Indeed such knowledge is essential in order to reliably measure and correctly interpret the resulting wave forms. Similarly, the potential risks of evoked potential measurement in terms of electric shock hazard can only be appreciated if one has some knowledge of basic safety requirements, both with respect to the equipment itself and, more importantly, to its use in clinical practice.

The clinical relevance of evoked potential measurements depends, in the first place, on the quality and reliability of the acquired evoked potential wave forms. Besides a good recording technique and artifact-free measurements such a requirement calls for standardization of various stimulation and recording parameters and the use of selected protocols for auditory, somatosensory and visual testing, depending on the clinical problem at hand.

The above summary reflects the contents of the present technical chapter. Emphasis has been put on practical issues, rather than theoretical background. For further technical reading references on selected topics are provided in each of the subsections. More general technical overviews can be found e.g. in the Handbook of Clinical Neurophysiology volumes 3, 4 and 8 or in Beagly, part I (1979).

2. GENERAL ASPECTS OF STIMULATION

As the physical appearance of the stimulus is totally different for the different modalities, and the choice of a particular form of stimulation depends heavily on the clinical problem at hand, we have decided to incorporate a discussion concerning these aspects in the corresponding subsequent chapters. The present section will deal with an issue which is common to all modalities, i.e. the timing of the stimulus. In particular, we will discuss the selection of the stimulus repetition rate and the choice between regular versus irregular (or "random") stimulation.

2.1. Stimulus Repetition Rate

The choice of a proper stimulus repetition rate depends, in the first place, on the part of the sensory system which has to be investigated. Peripheral, spinal and brainstem evoked potentials have relatively short recovery cycles. For these short latency evoked potentials one can, independent of the actual modality under study, therefore use relatively high stimulus repetition rates. However, some authors (e.g. Pratt et al., 1980, Stockard and Westmoreland, 1981) suggest to use rates no higher than approximately 10 Hz. The reason for this is that with increasing frequencies (up to about 100 Hz) the latency of the various evoked potential components gradually increases while their amplitude decreases (e.g. Salamy et al., 1978; Van Olphen et al., 1979; See Fig. 2.1).

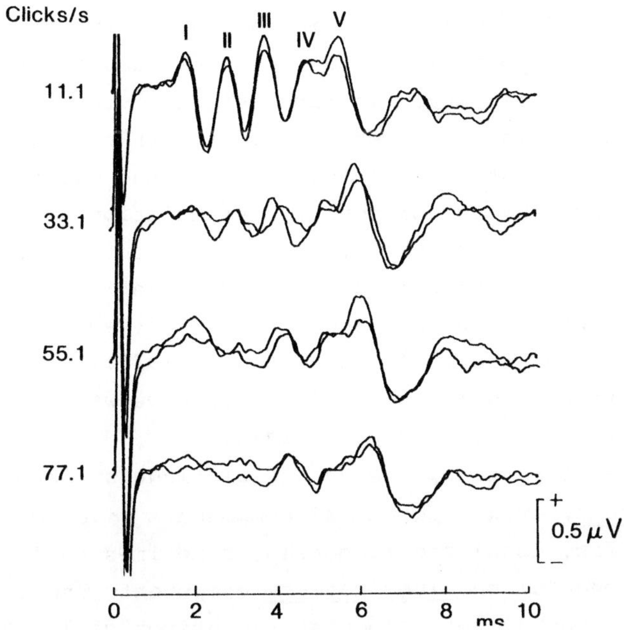

Fig. 2.1 Brainstem auditory evoked potentials (C_z to ipsilateral mastoid) from a normal subject in response to monaurally applied 80 dB, 100 us condensation clicks. Bandpass used: 150 Hz - 3 kHz. With increasing stimulus repetition rate the various waveform components show increased latencies and reduced amplitudes. Duplicate recordings have been superimposed at each click rate to show consistency of the results.

Moreover, it has been demonstrated that the rate of change shows variability among individuals (Stockard et al., 1978). Because of this variability it is important always to use a standard repetition rate. Even better is the use of two distinctly different (i.e. low and high) rates, since some abnormalities may only be revealed at higher repetition rates (Pratt et al., 1981; Yagi and Kaga, 1981). Integer fractions or multiples of the mains frequency (50 or 60 Hz) should be avoided to eliminate the possibility of stimulus locked mains interference.

When measuring cortical evoked potentials the notions transient versus steady-state stimulation come into view. The recording of transient evoked potentials requires that subsequent stimuli are sufficiently spaced apart in time such that the generating system (i.e. the brain) is given time to return to its resting state in between the stimuli. In contrast, steady-state evoked potentials involve the use of periodic stimulation at relatively high frequencies such that the brain does, intentionally, not return to its resting state. These evoked potentials can be analysed and quantified using frequency domain techniques (see e.g. Spekreyse et al., 1977 and Regan, 1977), but as these methods are rarely used in clinical practice we will confine ourselves to transient evoked potentials.

For this latter type then, one should distinguish between short and middle latency evoked potential components (i.e. cortical, stimulus specific, localized responses, sometimes called "primary" components) and long latency components (nonstimulus specific, more generalized responses, sometimes called "secondary" components).

To adequately measure short and middle latency cortical components one can use similar or slightly lower repetition rates as in measuring subcortical components, i.e. approximately 5 - 10 Hz. At higher stimulation rates similar changes in latency but more pronounced changes in amplitude take place as compared to subcortical components. (Fig.2.2).

On the other hand the measurement of long latency components requires significantly lower stimulus repetition rates, 1 Hz being approximately the upper limit (Fig. 2.3). In fact for these evoked potentials the recovery cycle is of a much longer duration (about 5 to 10 seconds) than the duration of the evoked potential (about 0.5 s) itself (Allison, 1962; Rothman et al., 1970; Woods et al., 1980), so that theoretically the stimulus repetition rate should be lower than approximately 0.1 to 0.2 Hz. Higher repetition rates cause an amplitude diminution of the late components, while a latency increase has been described as well (Stockard et al., 1979). However, in practice a rate of 0.1 Hz would lead to unacceptably long recording sessions, so that often higher rates, in combination with a certain random fluctuation in the stimulus interval time, are used.

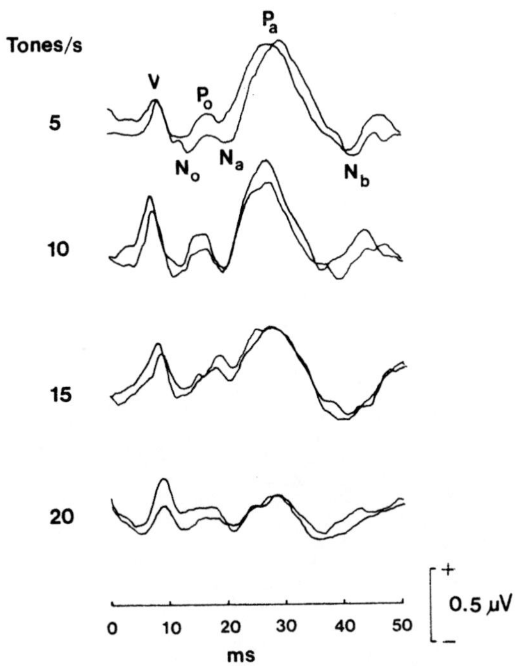

Fig. 2.2 Middle latency auditory evoked potentials (C_Z to ipsilateral mastoid) from a normal subject in response to monaurally applied 75 dB, 1 kHz, 10 ms, 3 ms rise-fall time tone bursts. The bandpass used was 10-250 Hz. These examples illustrate that the amplitude of middle latency components rapidly decreases as the stimulus repetition rate exceeds approximately 10 Hz.

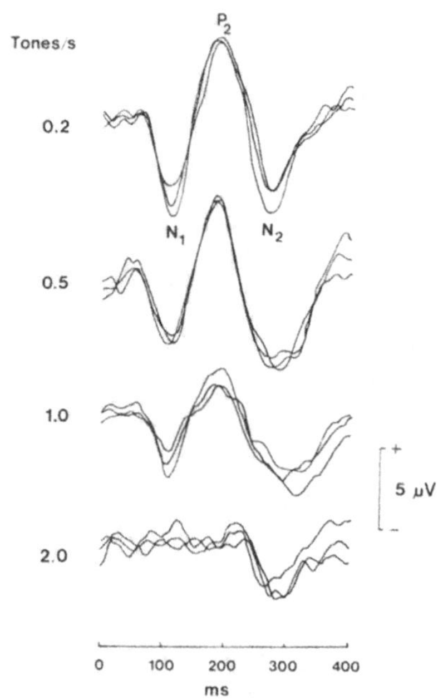

Fig. 2.3 Long latency auditory evoked potentials (C_z to ipsilateral mastoid) from a normal subject in response to monaurally applied 55 dB, 1 kHz, 35 ms, 3 ms rise-fall time tone bursts. Bandpass used: 1-100 Hz. At each stimulus repetition rate three replicate recordings were obtained. Note that, already at a rate of 1 Hz, the amplitude of the various waveform components is significantly affected.

2.2 Regular versus Irregular Stimulation

Apparently there are several misconceptions with respect to the advantages and disadvantages of regular versus random stimulation which seem, at least partly, to be due to the fact that older handbooks (such as the Handbook on Electroencephalography and Clinical Neurophysiology, Volume 8) stress the importance of choosing a proper stimulus timing for optimal background suppression rather than pointing out the effects of such timing on the evoked potential itself.

For irregular, or random, stimulation subsequent stimuli appear, within certain limits, at random time interval (Fig. 2.4).

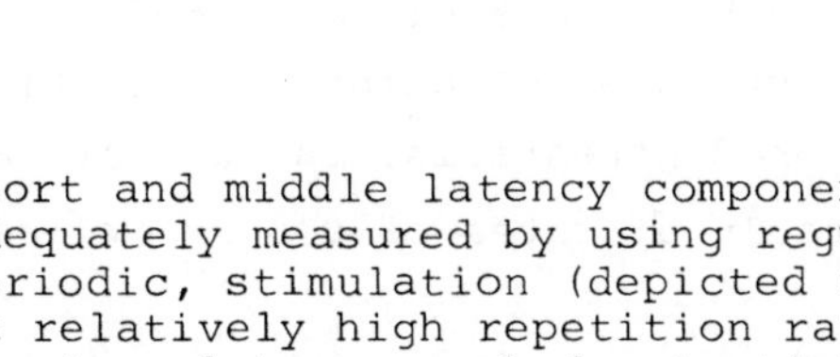

Fig. 2.4 a: Short and middle latency components can be adequately measured by using regular, or periodic, stimulation (depicted as arrows) at relatively high repetition rates.
b: For long latency evoked potentials irregular, or random, stimulation at low mean rates is the method of choice.

Usually, a range of variation of 50% of the mean stimulus interval is used (e.g. 2 $\pm$ 0.5 s).
This type of stimulation has certain advantages when recording long latency cortical components. Due to the random fluctuation in stimulus interval, the temporal uncertainty of the stimulus increases (Jackson and Barber, 1980). It has been suggested that this factor is responsible for a diminution in habituation

which means, in turn, a shortening of the recovery cycle (Wastell, 1980). Consequently this implies that random stimulation allows for higher (mean) repetition rates than regular stimulation. Moreover, random stimulation has favorable properties related to the suppression of background EEG if that background has a rhythmic character, such as the alpha rhythm (Ruchkin, 1965).

On the contrary it appears that when recording peripheral, spinal or subcortical components regular stimulation is the method of choice. As stated previously, the latency and amplitude of the various wave form components depend to some extent on the stimulation repetition rate. At short mean stimulus intervals, (i.e. high repetition rates) the random variation might therefore introduce a certain "jitter" (i.e. variability in latency) causing a smearing-out effect of the various components. Although these effects are expected to be small, it appears that, in these cases, random stimulation is an unnecessary expense, in view of the fact that early components do not show habituation.

In summary peripheral and central short and middle latency evoked potentials can be adequately elicited by using regular stimulation, the exact rate not being very critical. For long latency cortical evoked potentials, the use of random (irregular) stimulation, preferably at rates of 0,5 Hz or less, is indicated.

3. RECORDING THE ELECTRICAL ACTIVITY

3.1 Recording electrodes

The large majority of evoked potential recordings is performed non-invasively, by means of surface or skin electrodes. These electrodes are basically the same as the ones used in electro-encephalography and usually consist of small round cups of silver, tin or stainless steel. In auditory and visual evoked potential recordings, a similar electrode is used as ground electrode, but for somatosensory evoked potentials where the ground electrode is placed on the same limb which the stimulus is applied, different configurations are used. One possibility is the use of a relatively large metal plate, but we have found a wrap-around fabric covered lead electrode soaked in a saline solution before use, to be the most effective ground electrode.

It is important to degrease and lightly abrade the skin (e.g. with alcohol) and to apply electrode paste before or after attaching the electrodes. Such a procedure will reduce the electrode impedance considerably (to a few Kohms).

Undoubtedly, electrodes form a very critical link in evoked potential recordings and they should therefore be carefully applied. A thorough cleaning after each recording, periodic inspection and chlorization, in the case of silver-chloride electrodes, are essential prerequisites to keep the electrode impedance low and to obtain recordings free of artifact and interference.For the same reason electrodes of questionable integrety, which form the major source of intermittent artifacts and unreliable data, should be discarded.

3.2 Recording amplifiers

The amplifiers used in evoked potential systems are, without exception, of the differential type. This means that in the ideal case only the difference of the signals at both inputs is amplified and the common component cancelled. In practice, total cancellation is impossible because there will always be some imbalance at the two amplifiers inputs, which results in a limited so-called Common Mode Rejection Ratio (CMRR). With a high CMRR, effective suppression of interference signals can be achieved. This is a consequence of the fact that most of the sources of interference (such as the power distribution system) are relatively distant from the recording electrodes, and therefore cause fairly equal voltage variations at the two electrodes. Sometimes also nearby sources of interference (e.g. a stimulus artifact or muscle activity) can be suppressed by a proper positioning of the ground electrode.

Apart from a high common mode rejection ratio, other important requirements of the amplifier include a sufficiently high input impedance, in order to avoid distortion and loss of amplitude (Geddes, 1975) and a low noise level (in the order of a few microvolts peak-to-peak for a bandwidth between 10 Hz and 3 kHz). When modern amplifier techniques are employed these requirements do not pose any significant problems. Several commercially available evoked potential instruments have so-called floating, or isolated, amplifiers. This implies that the inputs, including amplifier ground, are galvanically completely isolated from the main part of the equipment. This feature provides additional protection against shock hazard in the possible event of an equipment defect.

3.3 Analog filtering: preliminaries

A point of major concern in evoked potential recordings is that of filtering, since this may have a significant impact on the resulting wave forms and, consequently, on the interpretation of the findings (see, for example, Rossini et al., 1981 and Scherg, 1982).

If we denote the evoked potential that we wish to measure by the signal and the unwanted disturbing background by the noise, then we may say that the main problem in evoked potential recordings is that of a low "signal-to-noise ratio" (SNR). Essentially, there are two complimentary ways for improving the SNR, namely averaging (see subsection 4.2) and filtering.

Simply stated, the objective of filtering is to remove as much of the noise contribution as possible, without any significant distortion of the signal. This can be done by suppressing those frequency regions where there is noise only and to pass other frequency regions where signal components (as well as noise) are present. For example, in brainstem auditory evoked potentials there are no significant signal frequencies below 30 Hz. But in that frequency range there is a lot of noise, namely the spontaneous EEG. That activity can be effectively removed by filtering without significantly affecting the evoked potential itself.

According to their function, filters can be classified in two basic types, namely lowpass filters, which pass low frequencies and reject high ones vs. highpass filters which have the opposite characteristic. From these basic types one can derive bandpass filters, which pass a specific frequency range and their counterpart: i.e. bandstop filters. So called "notch filters", used to reject the line frequency (50 or 60 Hz), are, in fact, an extreme form of a bandstop filter.

In evoked potential recordings one always deals with bandpass filters, which are in essence a series connection of a lowpass and a highpass filter (Fig. 2.5).

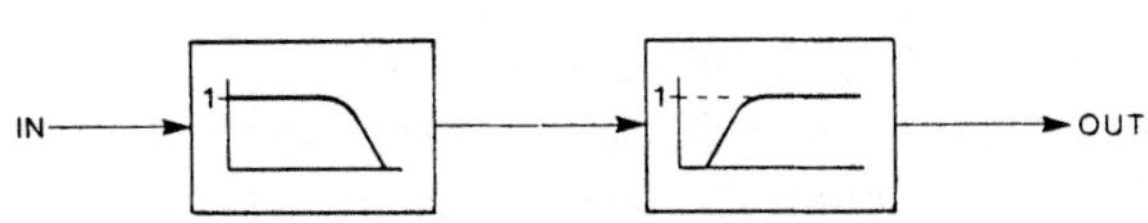

Fig. 2.5 In evoked potential recording one always deals with bandpass filters. These are made up by a lowpass and a highpass filter in series. Note that the lowpass filter has its cut-off frequency in the high frequency region and vice versa for the highpass filter.

Note that the cut-off frequency of the lowpass filter is in the high frequency range, while the cut-off frequency of the highpass filter is in the low frequencies, a point that often causes confusion. Therefore, a different terminology is sometimes used, e.g. "high bandpass" and "low bandpass".

The cut-off frequencies are usually specified in terms of the -3 dB points. This custom stems from engineering practice where filter attenuation, as well as amplifier gain, are oftentimes expressed in decibels (dB). For that purpose one computes the logarithm of the ratio of the output (V_o) and input voltage (V_i) and multiplies this by 20 (i.e. $20\ {}^{10}\log \frac{V_o}{V_i}$). At the -3 dB points the attenuation is 0.707 ($= 1/\sqrt{2}$) with respect to the nominal passband.

For what follows it will be important to notice that the frequency response curve of a filter consists of an amplitude response curve, which describes the amplitude attenuation as a function of frequency and a phase response curve, which describes the phase difference between the filter input and output as a function of frequency. For conventional analog filters, amplitude and phase response curves are connected, the precise relation being dependent on the filter design. A general consequence thereof is that, when using analog filters, it is impossible to obtain a bandwidth reduction without causing phase shifts as well.

3.4 Analog filtering of evoked potentials

There are two problems which make filtering of evoked potentials a less trivial matter than it may seem at first sight. In the first place no filter is ideal in the sense that its amplitude response has a sharp transition between the frequency region where it passes vs. where it rejects all frequencies. In practice there always is a region where the signals become gradually more attenuated outside the passband. A similar argument holds true for the frequency content of the evoked potential wave form, which, as a matter of fact, is also not confined to a sharply defined frequency region.

Therefore, the actual choice of the filter cut-off frequencies always remains a compromise between effective background suppression and possible wave form distortion, for which no dogmatic rules can be given (see also Doyle and Hyde, 1981a).

Secondly, and more importantly, not only the amplitude response should be considered, but the filter's phase response as well. In fact, the phase response may have significant influence, i.e. introduce considerable phase shifts at frequencies where the amplitude response is still flat. These phase shifts result in turn in shifts in the time of occurrence (latency) of peaks and troughs in the wave form (see e.g. Boston and Ainslie, 1980). Since latency measures play a key role in the interpretation of evoked potentials (see subsections 6.1 and 10.3) one should be well aware of these effects. From theoretical considerations it follows that an <u>increase</u> in either the low or high bandpass results in a <u>decrease</u> in latencies, while a decrease in either cut-off frequencies causes an <u>increase</u> in latencies. The precise latency shifts depend on a number of factors, among which are the high and low cut-off frequencies. However, of similar importance is the type of frequency response of the filter being employed (Doyle and Hyde,1981b). This point cannot be sufficiently emphasized, especially because it is wrongly ignored in the majority of evoked potential publications.

The filters currently employed in evoked potential equipment mostly have frequency response curves of either the simple RC, Bessel of Butterworth type (Fig. 2.6), each having its own specific advantages and disadvantages. Each filter is further specified by its "order" (Fig. 2.7), which determines the steepness of the slope of the amplitude response curve outside the pass band band and also determines the phase shifts between input and output (see, for example, Stapells and Picton, 1981). For simplicity, we assume here that phase shifts simply imply an overall time shift of the entire wave form, affecting all components in a similar manner. For some filters, however, such a simplification no longer holds true when the order of the filter is increased, so that the wave form may become distorted.

A detailed treatment of the entire filter subject is far beyond the scope of this chapter, but the important point to make is that specification of the cut-off frequencies of a bandpass filter alone (e.g. in terms of -3dB points) does neither fully specify its amplitude nor its phase response, and thus the latency shifts introduced by that filter.

Fig. 2.6 (see next page)
Amplitude response curves (top figure) and filter responses to a calibration pulse (lower figure a-d) for four different filter types: a. RC filter; b. Bessel filter; c. Butterworth filter; d. Cosine bell digital filter (see section 5.1). As can be seen in the top figure, all filters have the same cut-off frequency f_c, but the shape of their amplitude response curves near the edge of the pass band differs considerably and so do the filter responses to the calibration pulse (the pulse itself is shown by the heavy line in each of the lower figures).
COMMENT:

- a. The simple RC filter does not show any overshoot in response to an input step, but it has a less flat response in the pass band and suppression of the frequencies just beyond the cut-off frequency is less effective due to the moderate slope of the amplitude response curve.
- b. The Bessel filter does better with regard to the latter aspect and shows virtually no overshoot in the time domain.
- c. The amplitude response curve of the Butterworth filter offers a good compromise between flatness of the response in the pass band and suppression beyond the cut-off frequency, but this is at the expense of considerable overshoot in the time domain.
- d. In a later section (5.1) digital filters will be dealt with. The actual filter shown here provides a fair compromise between time and frequency domain characteristics. More importantly, this filter does not show any phase lag, or latency shift, as opposed to the other filters. Note, however, that the digital filter is responding already before the actual calibration pulse is applied.

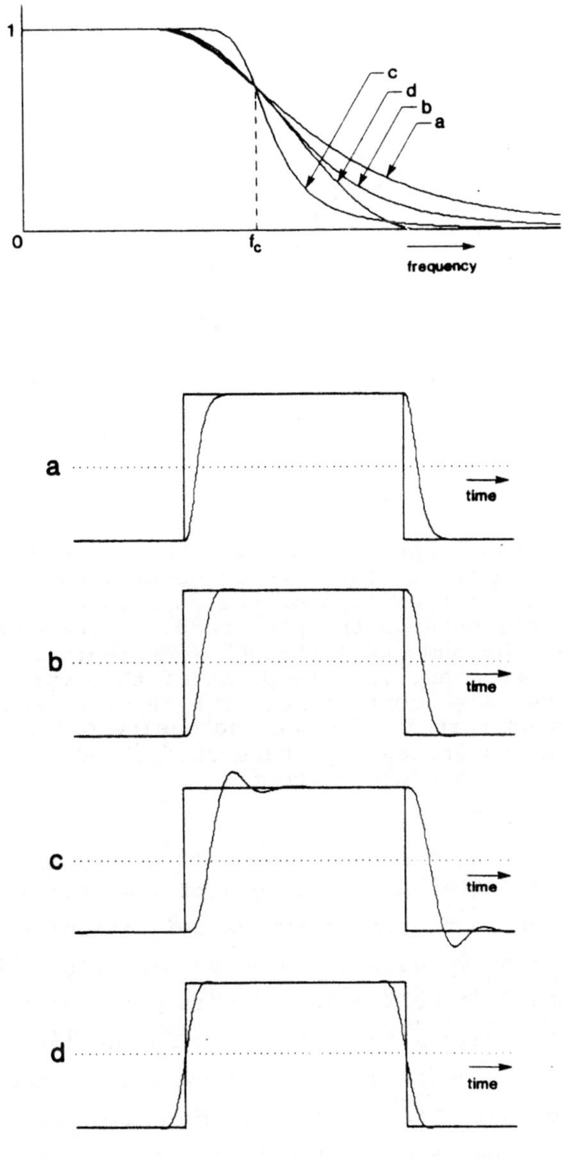

Fig. 2.6

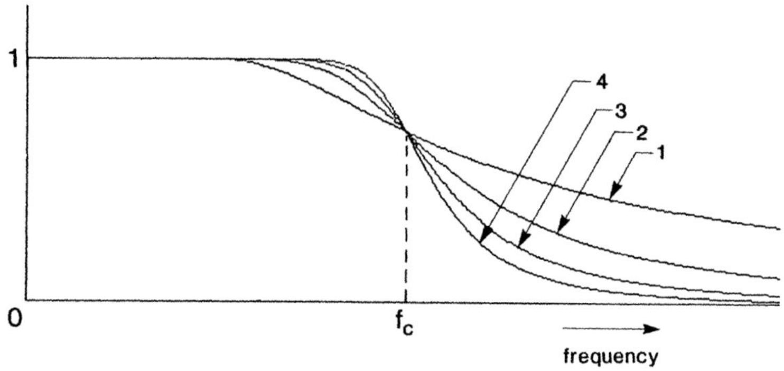

Fig. 2.7 An important aspect of a filter is its order, or the number of poles, which determines the steepness of the slope of the amplitude response curve well outside the pass band. In the above figure, the numbers 1 through 4 indicate the number of poles. The steepness of the amplitude response curve corresponds to 6 dB/octave/pole. Not shown here is that an increasing order also causes increasing phase shifts and associated waveform distortion.

In particular this latter aspect is one of the most important reasons (besides an inadequate stimulus definition) for the variability in normative data as found by different laboratories. Therefore one should be well aware of this point when comparing one's own normative values to those found in the literature. If large deviations are found it is worthwhile to experimentally verify the latency shift for some relevant filter settings, e.g. using simple sine waves at the filter input. In a more complex instrumentation set-up one should consider the possible presence of several filters in series (Desmedt et al., 1974), which may cause unexpected cumulative filter effects (Fig. 2.6).

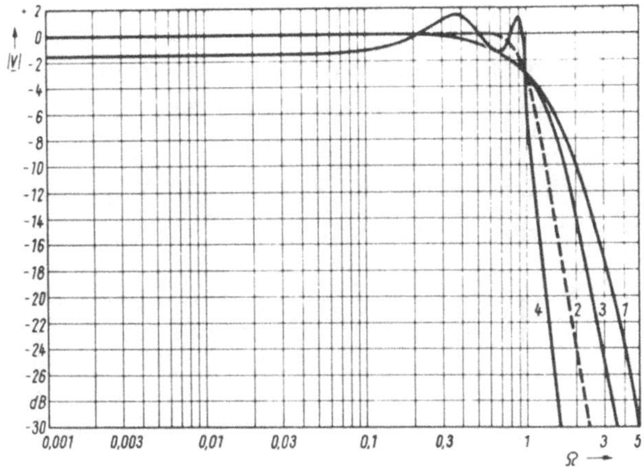

Fig. 2.8 In complex evoked potential systems where several, sometimes unnoticed, filters may be linked in series, the effective overall bandwidth of the system is likely to be reduced well beyond the bandwidth of each component individually, due to the cumulative nature of the filtering process. Although in the above example the bandwidth of the EEG amplifiers (1) exceeds that of the input filters for the computer (3), it nevertheless reduces the overall bandwidth of the total system (4). The bandwidth of the FM recorder (2) plays a minor role in this case, since it is much larger than that of the other components.

4. ACQUISITION AND PROCESSING

As mentioned in the foregoing, the main problem in evoked potential measurements lies in the fact that these potentials are usually very small as compared to the background activity (Table 2.1)

Since filtering usually does not suffice for extracting the desired wave form from the background activity, additional signal processing techniques are necessary, averaging being the most commonly applied method.

Present-day averagers operate without exception on a digitial basis and for that reason the incoming analog signals have to be sampled and digitized, a process known as analog-to-digital (AD) conversion. Furthermore, modern averagers include facilities for artifact detection and rejection. The present section is intended to make the non-technical reader familiar with the basic principles of these technicalities.

EVOKED POTENTIAL TYPE	AMPLITUDE (μV p-p)	DURATION (ms)	BANDWIDTH (Hz)
AUDITORY			
Brainstem	0,5 - 1	8	30 - 3000
middle latency	1 - 5	50	30 - 200
long latency	10 - 20	300	0,1- 70
VISUAL			
early (oscillatory components	1 - 2	50	50 - 250
middle and long latency	10 - 20	300	0,1- 70
SOMATOSENSORY			
short and middle latency median nerve	1 - 5	30	30 - 1500
short and middle latency sural nerve	1 - 5	60	5 - 250
long latency	10 - 30	300	0,1- 70
SPONTANEOUS EEG	100	-	0,1- 70

Table. 2.1 Global indications of normal amplitude range, duration and bandwidth of various evoked potentials

4.1 Analog-to-digitial conversion.

The first step in making an analog signal suitable for digital processing, is that of sampling the signal at equidistant time intervals. The magnitude at subsequent sample moments is then converted into a digital code and stored in an ordered sequence of "words". The rank of a particular word in the sequence reflects the time at which the corresponding sample was taken, while the content of that word gives the actual amplitude of the original signal at that point (Fig. 2.9).

From the theoretical point of view, the sampling interval

Δt, or inversely, the sampling rate $f_s = 1/\Delta t$ should be such that f_s equals at least twice the highest frequency which is still present in the signal in order not to loose any information and to avoid the problem of so-called aliasing (Fig.2.10).

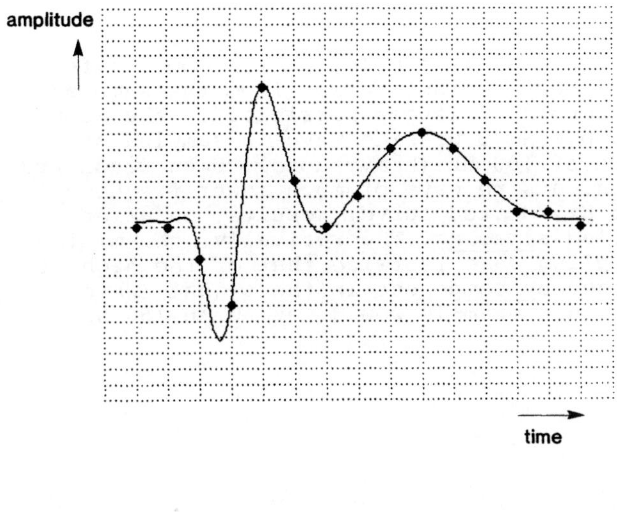

Fig. 2.9 Schematic illustration of the analog to digital conversion process. At equidistant points in time a signal sample is taken and the actual amplitude converted to the closest nearby digital value (heavy dots). The precision by which the converted values reflect the original waveform amplitudes depends on the resolution (number of bits) of the analog to digital converter.

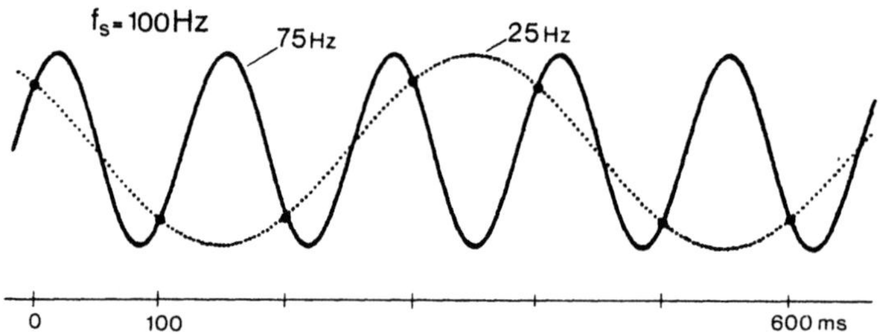

Fig. 2.10 If a signal is sampled at too low a rate, a correct reconstruction from the sampled values is no longer possible. In the above example a 75 Hz sine wave is sampled at a rate of 100 Hz (heavy dots). Upon reconstruction, a sine wave of much lower frequency (25 Hz, dotted curve) results, an effect called aliasing. Theoretically, the minimum sampling rate is twice that of the highest frequency component in the signal (which for this example would have been 150 Hz.

However, in practice, where one is interested in a reasonable wave shape reproduction by simply displaying consecutive sample points a sample rate of at least five times the highest signal frequency is more appropriate (Fig. 2.11).

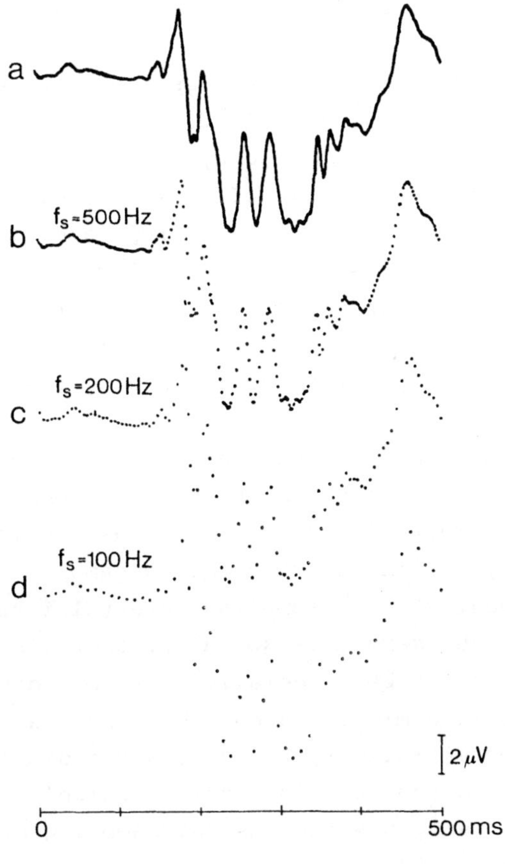

Fig. 2.11 For a reasonable visual reconstruction of sampled data on a raster type display it is necessary to sample the data much faster than would be necessary from the theoretical reconstruction viewpoint. In the above example the frequency content of the evoked potential in (a) extends to approximately 100 Hz so that a sampling frequency of at least 500 Hz is required to allow a reasonable waveform display (b). Theoretically, a 200 Hz sampling frequency would have been sufficient in order not to lose any information (c). At a sampling frequency of 100 Hz some of the higher frequency components clearly get lost (d).

As an example one may consider brainstem auditory evoked potentials where the highest relevant frequency is usually taken to be 3 kHz. The sampling frequency should thus be at least 15 kHz.

A related point of concern is the desired time resolution for measuring the latency of individual wave form components, a factor which is also determined by the sampling interval Δt. In general, when the above rule of thumb is followed, the associated time resolution will be appropriate. In the above example this resolution is $1/15 \cdot 10^3 \approx 0,07$ ms.

Although the sampling interval thus appears to be a relevant parameter, in most equipment its actual value cannot be explicitly chosen. Rather it is determined by the choice of the sweep speed or time base T and the number of available sample points N, i.e. the sampling interval $\Delta t = T/N$. For instance, when measuring long latency components of a somatosensory evoked potential, where T may be as long as 500 ms we have, assuming a memory length of 500 words, a sample interval $\Delta t = 1$ ms. Evidently, this sampling interval is inadequate for measuring at the same time spinal or short latency components. One should be especially aware of this problem when either short and middle latency or middle and long latency evoked potentials are to be measured simultaneously. When computer-based instrumentation is available this problem can be solved by using non-equidistant sampling schemes (Michelini et al., 1982).

Of much less importance than the sampling interval is the actual number of quantization levels or "bits", by which the magnitude of the signals is represented. Usually, conversion takes place with 8, 10 or 12 bits, which correspond to a data resolution of 1:256 ($=2^8$); 1:1024 ($=2^{10}$) or 1:4096 ($=2^{12}$). Due to the fact that an evoked potential is never a single recording, but always an average of several recordings, possible errors in quantization of the signal tend to average out. In fact, it can be shown that when significant background activity ("noise") is present and consequently a large number of sweeps has to be averaged, only a few bits suffice for obtaining a reliable average (de Boer, 1979). Of course it is thereby assumed that the incoming signals are properly amplified such that the full range of the AD converter is effectively utilized.

4.2 Signal averaging

The basic principle underlying the method of averaging is that the repeated presentation of an identical stimulus will each time lead to essentially the same evoked potential (the signal), which can be thought of to be additive to the disturbing background activity (the noise) which is totally unrelated to that stimulus. The desired result is obtained by averaging the entire collection of individual responses, whereby the underlying evoked potential will remain unchanged, and the background activity, occurring with random amplitudes and polarity will tend to average out. (Fig.2.12).

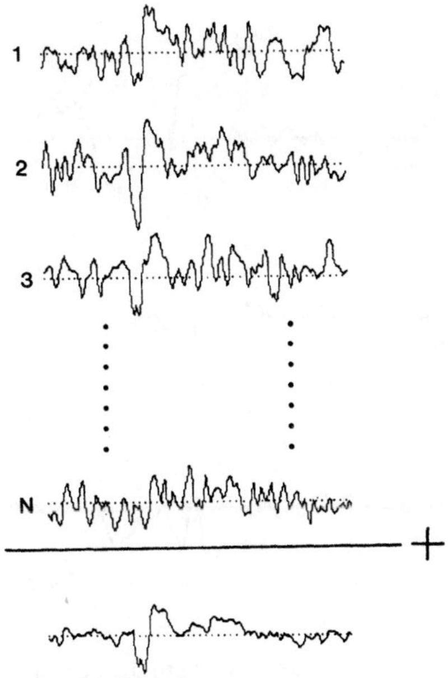

Fig.2.12 Schematic illustration of the averaging process. N represents the total number of sweeps averaged.

It can be proved that, under certain conditions, the signal-to-noise ratio (SNR) improves by the square root of the number of responses that are averaged. For example, when 100 responses have been averaged, the amplitude of the remaining noise has been reduced by a factor of ten as compared to the noise in individual responses.

In practice, the improvement in signal-to-noise ratio through averaging is limited by several factors. Firstly, as pointed out in the above, it is essential that upon repeated stimulation the evoked potential wave form remains the same. Particularly when recording cortical evoked potentials this requirement is difficult to maintain due to various physiological factors such as attention, habituation etc. which influence the response. Such influences may cause a variability in the time of occurrence of the evoked potential relative to the stimulus (so-called "latency-jitter"), in the amplitude of the various components and even in the wave form morphology (Fig.2.13).

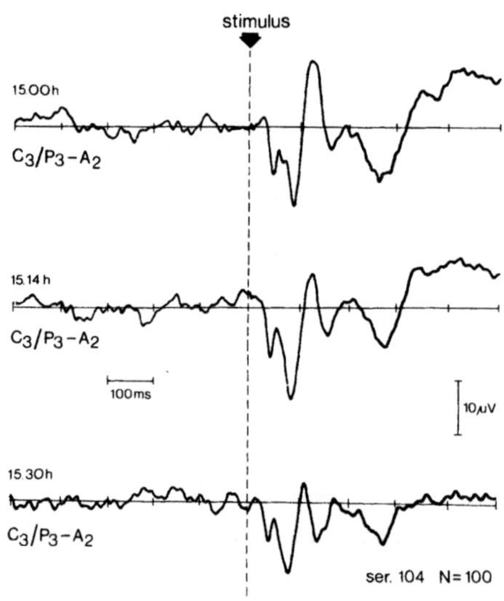

Fig. 2.13 Example of the wave form variability that may occur in subsequently recorded and averaged evoked potentials (median nerve somatosensory evoked potential). The most prominent effect is usually a diminution in the amplitude of the various wave form components as recording time proceeds.

In the second place, the square root law for SNR improvement only applies when the background activity consists of "pure" (uncorrelated) and stationary noise. On the other hand, ongoing EEG activity, in particular alpha rhythm, and electromyographic activity may show a distinct frequency or time structure such that averaging is much less effective in these cases.

As a consequence of the above factors, it is no uncommon finding in practice that averaging over long periods of time degrades, rather than further enhances the averaged evoked potential wave form. Particularly under these circumstances it is advisable to superimpose a number of duplicate recordings, each of a lower number of sweeps, instead of making only one average of a very large number of sweeps (see subsection 10.2). In addition more advanced processing methods may prove valuable in these cases (see section 5.3).

Another problem is the possible occurrence of spurious artifacts, which will be dealt with in the next subsection.

4.3 Artifact rejection

Artifact detection and rejection facilities are nowadays considered to be indispensable in evoked potential work. Artifact rejection is especially important when relatively few responses are averaged, such as e.g. in visual evoked potential recordings. In these cases a single large artifact may easily remain visible in the averaged wave form. In its simplest form, an artifact detection circuit signals the condition that the absolute magnitude of an incoming response exceeds a certain threshold, which is normally set close to the upper and lower range of the analog-to-digital converter. If this condition occurs that response is rejected, i.e. not included in the averaging procedure. In more sophisticated versions of artifact detection one can specify a minimal time interval for which the threshold must be exceeded (in order to avoid that a single sharp spike in the response causes that response to be rejected), or one can define a certain time window around the moment of stimulation to be excluded from artifact detection. The latter option is, in fact, extremely useful in evoked potential work since the stimulus artifact may easily be an order of magnitude larger than the response recorded, particularly in somatosensory and auditory stimulation.

5. FILTERING OF EVOKED POTENTIALS REVISITED

The fact that, after previous processing methods, the (averaged) evoked potential wave forms are available in digital form opens up new possibilities for further improvement of the "signal-to-noise ratio", in the form of digital filtering or the essentially equivalent method of "smoothing". These operations can either be fixed (pre-programmed) or data adaptive, as will be further elaborated on in the present section.

5.1 Digital filtering

The main characteristic of digital filters is that they operate on sampled data (see subsection 4.1). In principle, any type of analog filter can also be realized in a digital equivalent. However, in addition digital filters have, unlike analog filters, the interesting property that the amplitude and phase response curves can be designed independent from one another. In fact, it is thus possible, and widely practiced, to design digital filters according to a predetermined amplitude response curve, while maintaining zero phase response over the entire frequency range. This has the great advantage that the filtering operation can be performed without introducing any latency shifts in the filtered wave form (cf. Fig.2.6d).

(As an aside it may be noted that the above property has not to do with the sampled or digitized character of the data but rather with the fact that the data is filtered in retrospect, instead of in real time. Application of a zero phase filter requires the processing, not only of past input data, but also of input data that occur conceptually later than the actual output of that filter and hence such a filter is not realizable in real time).

In view of the clear advantages of digital filtering vs. analog filtering, one could question why that method is not more widespread used. One reason for this is that there always remains a necessity for analog lowpass filtering, so as to limit the frequency range prior to the sampling process in order to avoid the problem of aliasing (see subsection 4.1). Another, practical, reason lies in the more complicated instrumentation needed for the realization of digital filters. However, digital filters may be easily realized in the form of a computer program and as evoked potential equipment becomes more and more based on micro and minicomputer systems the use of digital filters will certainly become much more widespread in the near future (see e.g. Møller, 1983).

Finally, it should be noted that there is no difference in applying a filter to each of the individual responses prior to averaging or, alternatively, once to the final average. Averaging and filtering represent commutative operations, which implies that their order of application to the data can be interchanged. This is an important property, as it enables the final evoked potential wave form to the studied with various different filters without the need to acquire a new ensemble of responses each time.

5.2 Wave form smoothing

Wave form smoothing is a special case of digital lowpass filtering, available on several modern evoked potential machines. Usually, the smoothing is performed by weighing each data point with its nearest two neighbours (Fig.2.14), which caused fluctuations in nearby sample points ("high frequency noise") to flatten out. In the frequency domain this operation corresponds, simply stated, to filtering the data with a lowpass filter with a cut-off frequency corresponding to ¼ of the original sampling frequency. For instance, when a brainstem auditory evoked potential is acquired with a sampling frequency of 15 kHz, the above smoothing implies a lowpass filtering of 3.75 kHz. However, the amplitude response curve of this type of filtering is less flat in the passband than that of conventional analog filters, so that results of both types of filtering cannot be directly compared.

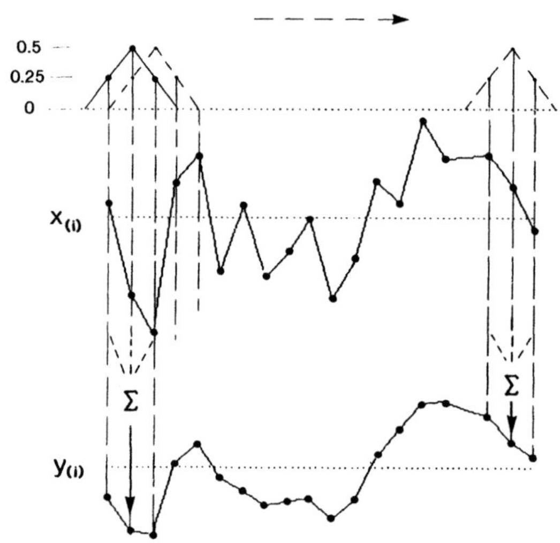

Fig. 2.14 Smoothing is essentially a lowpass filtering process whereby local fluctuations in the evoked potential wave form are flattened out by applying a moving average window. In the above example a three point "Hanning" window (top) with weights 1/4, 1/2, 1/4 is applied to the raw wave form x(i), resulting in the smoothed wave form y(i). Mathematically, each individual data point y(i) is formed by the weighted sum 1/4 x(i-1)+ 1/2 x(i) + 1/4 x(i+1).

Smoothing is a zero phase filtering and does therefore not affect the latency of the evoked potential components. The smoothing operation may be repeated a number of times to obtain more pronounced lowpass filtering (see e.g. Wastell, 1979), but the net effect diminishes on repeated application since the effective cut-off frequency does not reduce proportionally with the number of repetitions.

5.3 Adaptive digital filtering

In the foregoing, we discussed the use of digital filters with an *a priori* chosen amplitude response curve (and a zero phase curve). The use of such filters is based on (possibly implicit) knowledge about the global frequency contents of the acquired evoked potentials as well as that of the background noise. Clearly such filters cannot be really optimal for each particular case individually. A better method would be to establish the frequency content of the signal and noise for each separate evoked potential and accordingly devise a filter which is optimal for that particular case. This is exactly what happens in so-called "a posteriori"types of filtering, such as "Wiener" and "time-varying filtering". Briefly the "Wiener"filtering method proceeds as follows (Walter, 1969; de Weerd and Martens, 1978; de Weerd, 1981a). In addition to the normal average a so-called "alternate average" is obtained (see section 10.2), in which subsequent responses are alternately added and subtracted so that the signal cancels out and an estimate of the background noise and its frequency content can be obtained (cf. Wong and Bickford, 1980). By comparing the alternate average to the normal one the frequency content of the underlying evoked potential can also be estimated and a proper filter computed. In principle, this filter would then be the optimal filter for the particular evoked potential under study.

In practice, a refinement of this method appears to be necessary since the frequency content of an evoked potential varies considerably with the latency beyond the stimulus.

Typically, this variation is reflected in the time course of an evoked potential: from low amplitude, small duration (i.e. high frequency) short-latency components to high amplitude, large duration (i.e. low frequency) long-latency components. This observation has led to the development of the so-called a posteriori time-varying filtering technique (de Weerd, 1981b) which is able to provide substantial gains compared to averaging, either for improving the quality of the averaged evoked potential wave form (Fig.2.15) or, alternatively, for reducing the number of averaged responses while retaining approximately the same quality of the wave form.

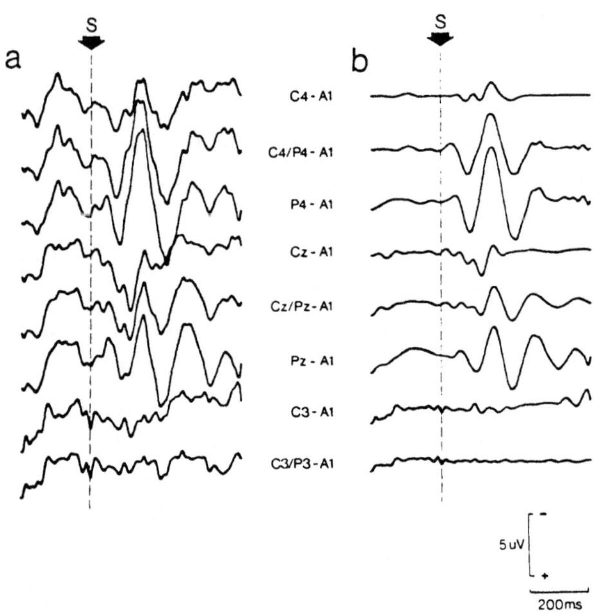

Fig.2.15 Comparison of normal averages of 128 sweeps (a) and their time-varying filtered counterparts (b) of long latency somatosensory evoked potentials from a seven year old male patient in response to electrical stimulation of the index finger. S indicates the presentation of the stimulus. Note the suppression of the background activity in the prestimulus interval and the enhancement of the wave forms in the poststimulus interval due to the time-varying filter procedure.

Of course the above techniques require, at least at the present time, extended computing facilities both in terms of computer memory as well as processing capability. However, with the current trend of dramatic improvements in micro-processor technology coupled with a steadily decreasing price, a more widespread application of these and similar techniques clearly come into view.

6. ANALYSIS AND DOCUMENTATION

6.1 Identification and labeling of components

Once an averaged evoked potential has been obtained the next step is to analyse its wave form. This is done by identifying various components, i.e. peaks and troughs in the wave form, from which both the latency and amplitude are determined. Generally only a selected number of peaks and troughs are measured, the actual choice being dependent on the type of evoked potential (see next chapters). It should perhaps be emphasized that, especially under pathological conditions, the identification of components is by no means a trivial matter and depends heavily on practical experience. Possible support may come from inspection of additional evoked potential wave forms from different channels and/or obtained under different stimulation or recording conditions. Of course it is thereby important to have exact knowledge of the electrode montage that was used as well as of the polarity of the evoked potential. Especially the latter aspect often gives rise to confusion in practice. For Auditory Brainstem potentials it is common practice to display vertex positivity upwards and for all other evoked potentials negativity on the active electrode upwards.

The latency of a particular wave form component is always determined as the time difference between the onset of the stimulus and the occurrence of a local maximum or minimum. In commercial equipment the latency can mostly be determined by using a marker or cursor which generates an intensified dot or a vertical line across the wave form displayed on the oscilloscope, and which is connected to either a mechanical or an electronical time display. When computer based instrumentation is available (semi) automatic peak detection algorithms may prove to be helpful (Gabriel et al., 1980; Fridman et al., 1982).

The amplitude of a wave form component can be measured either with respect to the "baseline" or "peak-to-peak" with respect to the nearest neighbouring component of opposite polarity. The baseline, in turn, can be defined in a variety of ways. A frequently employed method is to use technical zero, which is oftentimes implicitly defined as the midvalue of the A-D converter range. In adopting this method, one should be aware of the fact that the actual zero level usually shows some long term drift due to a variety of technical factors. Periodic inspection of the zero level (see section 7.2) by making up an average with the amplifier inputs shorted to ground is essential for retaining reliable amplitude measures in this case.

Another approach is to compute the mean value across the entire wave form and to define that value as the baseline. The disadvantage of this method is that the result largely depends on the time interval (time base) over which the mean is taken. In addition, the presence of stimulus artifacts, the tails of an electrical artifact in particular, makes this method unreliable.

Still another method is to use the mean value of the pre-stimulus interval (see section 10.2). This approach relies on the assumption that the stimuli are sufficiently spaced apart in time so that the pre-stimulus interval does not contain the tail of the response, evoked by the preceding stimulus. In any case it is necessary to have an understanding as to which of the above methods is used in a particular evoked potential instrument so that one is aware of any possible pitfalls.

Peak-to-peak amplitude measurements are less susceptible to the above mentioned problems and tend to provide more reproducible and reliable results. On the whole, however, amplitude measures do show a significantly larger variability than latency measures.

Unfortunately, the labeling of evoked potential components has not yet been internationally agreed upon. Historically, auditory brainstem potentials have been labeled with roman numerals (Jewett and Williston, 1971), while for the other evoked potentials the actual tendency is to label components according to their polarity and (normal) latency, such as P_{100}. However, several other nomenclatures are still in use and are, occasionally, even preferable, e.g. when the polarity of components changes with the location of the recording electrode, or when the maturation of evoked potentials is studied with associated variations in latency of the components.

6.2 Waveform plotting and documentation

Besides having a descriptive qualification of the recorded evoked potentials in terms of latencies and amplitudes of various components it is good practice to also document the acquired wave forms themselves. Not only does this provide the additional information of wave form morphology and possible time-relationships between various recorded channels, but it also allows visual appreciation of the quality of the recorded potentials in terms of remaining background activity and possible (stimulus) artifacts.

Usually, averaged evoked potentials are written out on an X-Y recorder. It is thereby essential that both the time and amplitude axes are accurately calibrated; either by using a standard write-out on graticule paper or by including calibration bars for time and amplitude axes along with the plotted wave forms.

Since an averaged evoked potential wave form is useless without complete knowledge of the stimulus and recording parameters such information should, in some way, be included with the wave forms. When using standard protocols, an indication of which protocol was used may suffice, but otherwise it is necessary to annotate the wave forms with essential parameters such as type, intensity, frequency and duration of the stimulus, and high and lowpass filter settings to mention some of the most important ones.

Furthermore basic data such as a patient identification, recording date, and a short descriptive statement about the course of the investigation and the patient's state should be included as well.

Some of the more sophisticated evoked potential machines have facilities for producing listings of all or part of the above parameters, which is an advantageous feature both in terms of efficiency and reliability of the results obtained.

6.3 Digital wave form storage

A drawback of storing evoked potential wave forms on paper only is the fact that re-evaluation and quantification in a later stage can only be performed using a ruler. On the other hand the storage of wave forms in some electrical manner may even allow for additional processing of the data in retrospect.

Since averaged evoked potential wave forms are generally represented in digital form the most obvious way of storage is on a digital medium such as flexible or hard disks, interfaced in some way to the averager. At present, the available possibilities range from small home-computers equipped with mini-floppies to systems where floppies or even so-called "Winchester"-disks form an integral part of the system.

For digital wave form storage too, it is important that the relevant stimulus, recording and patient parameters are stored and can be easily retrieved along with the wave forms.

7. EVOKED POTENTIAL INSTRUMENTATION

Thus far we have considered the various technical aspects of evoked potential measurements from a rather theoretical point of view. We will now discuss more practical aspects, starting from the equipment that is required.

7.1 Basic evoked potential system

The basic configuration of an evoked potential system consists of the following elements (Fig.2.16):

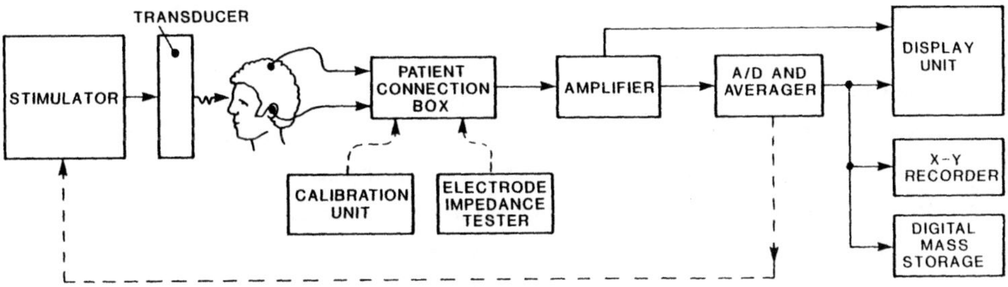

Fig. 2.16 Block diagram of an evoked potential system.

1. Stimulator and associated transducer.
   Depending on the specific modality to be investigated and the type of stimulus that is desired these may include: auditory stimulator and earphones; current stimulator and stimulus electrodes; stroboscope, light-emitting-diode (LED) goggles, slide projector or TV monitor;
2. Amplifiers, including recording electrodes.
   For evoked potential measurements at least two, and preferably four, amplifiers are required;
3. Averager, or any other hardware device which can perform the averaging operation, such as a micro- or minicomputer. The averager should have artifact detection and rejection facilities;
4. Display unit (e.g. oscilloscope) and system for documenting the final wave forms, e.g. XY-recorder, photo-camera or digital plotter;
5. Electrode impedance tester, based on AC measurement to circumvent the problem of electrode polarisation effects, and

6. Calibration unit. It is preferable to have a unit that can be plugged directly into the patient connection box so that the entire system integrity can be checked and proper calibration is guaranteed.

From the above summary it follows that evoked potential systems have a lot in common with electromyographic equipment and indeed instruments of the latter type are often employed for evoked potential investigations. However, facilities for electrode-impedance measurements, which must be considered absolutely essential, are usually not incorporated in such systems. Also, the available filter settings on EMG amplifiers sometimes deviate appreciably from what is needed in certain types of evoked potential recordings. These points need special consideration and instrumental modifications or adaptations may turn out to be necessary.

In addition to the above mentioned functional elements, more sophisticated evoked potential systems may include:

7. Cursor controls for on-line determining latency and amplitude of evoked potential wave form components;
8. Programmable stimulator, capable of delivering specific stimulus sequences and/or random (irregular) stimulus intervals;
9. Eight or more input channels, which allow the study of the topographic distribution of evoked potentials over the scalp;
10. Wave form manipulation facilities (such as smoothing, addition, subtraction);
11. Wave form annotation facilities (stimulus and recording parameters, administrative data etc.);
12. Digital storage media, such as floppy or hard disks and
13. Temperature unit and sensor. This unit is particularly important when measuring somatosensory evoked potentials (see chapter 5).

Several nowadays commercially available evoked potential systems include all of the above mentioned facilities and some have onboard or plug compatible micro- or minicomputer capability, thus providing flexible and powerful instrumentation. The latter type of equipment is usually fairly complicated, but provides, at the same time, possibilities for using completely automated protocols. In these protocols all stimulus and recording parameters are set under computer control, thus simplifying investigation procedures and eliminating a possible source of errors.

7.2 Periodic inspection and calibration

For proper operation of the equipment over extended periods of time periodic inspection and calibration is necessary. Although this requirement may seem rather trivial, the experience in practice is, time and time again, that such form of maintenance is usually deferred until the moment that a defect occurs.

As stated previously, it is primarily the electrodes that need special attention. The state of the electrode surface and the quality of the cable connection should be inspected on a regular basis and the entire electrode tested, which can be done by using a simple ohmmeter.

Inspection of the equipment typically will start by acquiring a dummy average (see subsection 10.2) with the electrode short-circuited. This enables a check on proper operation of the instrument, the base-line noise and possible DC drift. It is highly advisable to keep a log-book with each evoked potential machine where test results, including the recorded wave forms, are documented for periodic comparison Adjustments, repairs etc. should be noted in the book as well.

In modern equipment, necessary calibration is reduced to a minimum and usually does not extend beyond adjustments of amplifier gain and DC base-line and, sometimes, X and Y gain and offset of the XY recorder. As a rule, the manufacturer provides detailed guidelines for the procedures to be followed. The calibration unit should deliver a stable wave form of precisely known shape, duration and amplitude, triggered by the stimulus. Calibration units need regular inspection too, even though they should be an order of magnitude more accurate and stable than the equipment that they are applied to. In particular, when the calibration unit is battery powered, the state of that battery should be checked prior to performing any major recalibration of the equipment.

8. ELECTRICAL SAFETY

In order to assure safe evoked potential measurements it is necessary that the evoked potential equipment, the power circuit to which it is connected and the electrical environment of the investigation room comply with some basic safety standards. Furthermore, in case of a possible hazard, sufficient protection should prevent the patient or the equipment's operator to be unintentionally exposed to electrical shock. It is also important to be able to make a judicious use of the apparatus, according to some basic rules, in particular with respect to patient grounding and electrical stimulation. In the following, these aspects will be dealt with in more detail.

8.1 Safety Standards for Equipment

Electromedical equipment should comply with a number of strict requirements for safety and reliability. Even in case of a possible defect (such as an interrupted ground lead in the power cord), the apparatus must still provide sufficient protection against electrical shock hazard. Although mandatory safety standards still do not exist at present, the recommendations of the Underwriters Laboratories (in the U.S.A., Standard UL544) or the International Electrical Committee (in Europe, Standard IEC 601-1) are generally accepted. An important user's aspect of these standards is the criterion for maximally permissible leakage currents in medical apparatus during normal operating conditions and in the case of an equipment defect (a so-called "single fault condition"). Two different types of leakage current are of major concern.

The first type is the ground leakage current, which is normally shunted to ground by the ground lead. This current is mainly a consequence of the capacitive but also of the inductive coupling of the power line transformer to other parts of the apparatus. Normally this current presents no danger, unless for example the ground connection is interrupted. The IEC standard requires that the ground leakage current be less than 0.5 mA under normal operation and less than 5 mA in case of a single fault condition. The UL544 does not deal with the ground leakage current as a separate item.

The second type of leakage currents normally do not flow through the ground lead but through the patient or person operating the equipment. These are respectively the so-called patient and enclosure leakage currents, which can present a danger even under normal operating conditions. Unlike the UL544, the IEC standard distinguishes between general patient care equipment (categorized as class B or BF) and equipment which may come in direct conductive connection to the patient's heart (class CF equipment). According to the IEC standard, the maximally allowable patient and enclosure leakage currents for class B or BF equipment are 0.1 and 0.5 mA for normal operating and for a single fault condition respectively. The UL544 standard poses more severe requirements. Under a single fault condition the maximum allowable patient and enclosure leakage currents are 10 uA and 100 uA respectively.

Many of the commercially available evoked potential systems comply with these requirements. It is important to realize that purpose built evoked potential equipment centered around general purpose averagers, laboratory computers or micro-processors may easily exceed one or more of the above limits for permissible leakage currents by an order of magnitude. In those cases, the employment of additional precautions such as isolated, or "floating", preamplifiers (see section 3.2) and/or line isolation transformers (see next section) are required.

In any case the equipment should be inspected when put into use for the first time and periodically thereafter, especially on ground connections and leakage currents. A detailed description on the procedures for doing this can be found in the IEC 601-1 report.

8.2 Power line safety measures

Although medical equipment should, by its very nature, carry adequate electrical protection, the risk of electrocution is not imaginary in case of a gross defect. As an additional protection both for the patient and the operator, the installation of a "Ground Fault Circuit Interruptor"is therefore desirable. An even better method is to supply the entire laboratory from a power line isolation transformer provided with a " line isolation monitor" (see e.g. "Electricity in patient care facilities" by the National Fire Protection Association). A Ground Fault Circuit Interruptor detects any leakage current flowing in the ground circuit (see section 8.1) and automatically switches off the power in case this leakage current should rise above a preset limiting value (e.g. 5 mA). However, this type of security does not provide sufficient protection if the leakage current should unexpectedly flow through the patient; only the duration of the current flow will be limited, but the maximal intensity may be far above the safety norms (cf. section 8.1). It is therefore safer to employ an isolation transformer, which has a very slight capacitance between the primary and secondary windings due to a good shielding. An isolation transformer offers a complete galvanic separation from the "normal" power distribution system. The great advantage of this is that the leakage current remains under the safety margin even if the patient, e.g. as a result of a serious defect in the apparatus, should come into contact with one of the current-carrying wires of the (isolated) power circuit. This results from the fact that the secondary supply lines are completely "floating" with respect to ground. It is desirable to have isolated power circuits provided with a so-called "line isolation monitor", which essentially is a leakage current indicator. By this means leakage currents can be detected, which arise through defects in the apparatus or the electrical installation, carrying the risk that the patient could possibly come into contact with one of the current-carrying lines. These systems are designed such that even at very small leakage currents (of the order of magnitude of a few tens of uA) an alarm signal is set off. Isolation transformers should always be installed in the immediate neighbourhood of the laboratory, since it is important to have the secondary leads kept as short as possible.

Furthermore, the employment of an isolation transformer only makes sense if all the equipment in the relevant laboratory is supplied by the isolated power circuit. Wall receptacles connected to the "normal" power distribution system should therefore not be installed in the same room with those connected to the isolated power circuit.

Finally, a uniform "ground potential" throughout the laboratory is of great importance. This means that between the various grounding poles there should essentially be no difference in potential and that all the ground leads should be connected to a common point.

8.3 Safety in practice

The main risk associated with the leakage currents, as discussed in the above, is that of a possible current flow through the heart, causing e.g. ventricular fibrillation. This risk may be considerably reduced by observing some basic rules for proper grounding and, when measuring somatosensory evoked potentials, for proper electrical stimulation.

In the first place care must be taken that the patient is not doubly grounded, which may also occur accidentally e.g. when a grounded metal frame of a bed is touched or when the equipment is used in the operating room where other equipment has already been connected to the patient. Although under normal conditions this does not necessarily lead to hazardous situations, usually a troublesome interference voltage will result which is due to the flow of so-called potential equalizing currents in different ground leads.

Secondly, when recording somatosensory evoked potentials, using electrical stimulation, additional risks are introduced, due to the fact that electrical stimulators may induce lethal currents if improperly applied. The two stimulating electrodes should never be attached to different limbs and, unlike for auditory and visual evoked potentials where the ground electrode is placed near the recording electrodes on the head, the ground electrode must preferably be positioned in the neighbourhood of the stimulating electrodes and always on the same limb. This helps in keeping possible current fields local so that no significant current flow through the heart can occur. As an additional safety measure and to reduce stimulus artifact (see section 9.2) modern stimulators are always provided with an isolated output, i.e. an output which is galvanically separated from the ground.

Finally, extreme caution should be taken when connecting auxilliary equipment, and in particular non-medical equipment, to an evoked potential system, as this may violate the electrical safety of that system. Even when auxilliary equipment that meets the safety standards is connected it is important that all apparatus is grounded at the same point and supplied from the same phase of the power distribution system.

9. INTERFERENCE AND ARTIFACTS

If one realizes that problems with interference and artifacts can be quite troublesome when recording the spontaneous EEG, then it is not difficult to understand that these problems are a veritable plague when recording evoked potentials, whose amplitudes usually are an order or even two orders of magnitude smaller than the EEG. Moreover, these potentials are elicited by stimuli requiring relatively large driving currents, forming an additional and major source of artifact. Starting from basic constructional provisions, this section suggests various measures for eliminating the three most common types of artifacts, namely interference, stimulus artifacts and bioelectric artifacts. Stimulus artifacts are, by definition, time-related to the stimulus and this type of artifact is particularly cumbersome because it does not average out. Similarly, bioelectric artifacts may be stimulus-related, e.g. scalp muscle reflexes, although fortunately the majority is not. Finally, mains interference may become stimulus related if the stimulus repetition rate is improperly chosen (see section 2.1 for further details). Nonetheless as a general rule technical and bioelectric artifacts should not be "eliminated" by averaging, but should be removed as properly as possible prior to any evoked potential measurement.

9.1 General provisions for avoiding interference

A primary requirement for avoiding electrical interference is a careful planning of the location of an evoked potential lab inside a larger complex such as a hospital. Notorious sources of interference include diathermy and radiological apparatus, elevators and paging systems. In order to avoid the electro-magnetic fields and mains interference introduced by the above, the laboratory must be located sufficiently remote from such apparatus and must preferably be connected to a different phase of the power distribution system. Also, the wiring installation within the laboratory requires special attention in relation to the prevention of mains interference. In principle, interference of this nature can be introduced in two ways, namely capacitatively and inductively (interference as a result of variations in the mains voltage will not be considered here).

Capacitative transfer of mains interference arises because the body is a conductor, which possesses a certain capacitance both with respect to the power circuit and to ground. As a consequence a (very small) so-called "leakage current" flows through the body, which results in a voltage difference (or "hum") with respect to ground. This effect can be reduced considerably by grounding the patient. This is usually done with a ground electrode, which has, however, still an impedance of 1 kΩ or more so that this hum voltage never disappears completely.

Inductive transfer of mains interference can arise in the cables forming part of the electrode input and amplifier circuit (e.g. the electrode cables). If these cables form a loop (in the electrical sense) and are subjected to a varying magnetic field, for example of a power-line transformer, an interfering voltage is induced. An effective method for largely eliminating both types of interference is to enclose the power wiring in seamless, grounded metal tubes. In addition, it is important to provide all electrical appliances within the laboratory (including desk lamps etc.) with shielded and grounded power cables (this applies, of course, also to possible extension cords). It is also desirable to ground all appliances, lamp armatures etc. For the lighting of the laboratory ordinary lamp bulbs instead of fluorescent lights should be used. Equipment not directly needed for the investigation should preferably be removed or disconnected by removing the plug of the power cable from the wall receptacle. If all these precautions are taken, it will rarely indeed be necessary to provide complete shielding of the laboratory by means of a so-called Faraday Cage.

9.2 Measures for avoiding stimulus artifacts

Stimulus artifacts may occur in a variety of ways. Irrespective of the type of stimulation used, it is always advisable to keep the leads from the recording electrodes as remote as possible from the stimulating source and stimulator cables. Also, recording electrode leads should not be longer than strictly necessary.

In auditory brainstem recordings the main source of artifact is electromagnetic pick-up in the recording leads of the currents used to drive the headphones. This type of artifact can be significantly reduced by shielding of the earphones and the connecting cable. Also alternating click polarities (condensation vs. rarefaction clicks) may be used, so that the artifact cancels out during the averaging process (see, however, chapter 3).

Visual evoked potential recordings usually show the least problems with stimulus artifact, due to the fact that the stimulator is, in general, relatively remote. When using high intensity diffuse light flashes it may sometimes be necessary to place a grounded copper screen directly in front of the lamp to reduce electromagnetic radiation.

In somatosensory evoked potential recordings the situation is more complex. Here, stimulus artifact results from (i) the currents induced into the body, setting up a potential field and (ii) currents in the ground lead, due to capacitive coupling between the floating stimulator output and ground (see McGill et al., 1981 for a detailed discussion). To reduce these effects it is important to (i) use a close spacing between the stimulating electrodes, which keeps the current field local and reduces the extent of the potential field; (ii) place the ground electrode relatively close to the recording electrodes such that the common mode components due to the above currents remains small; (iii) if at all possible, search for an optimal position of the anode (keeping the stimulating cathode in place), so as to minimize the potential gradient across the recording electrodes. Of course carefully applied electrodes having a low skin-to-electrode impedance remain the most basic requirement for low stimulus artifacts.

9.3 Bioelectric and other artifacts

In evoked potential work one has to cope with artifacts due to EEG and EKG, eye blinks, eye movements and muscle activity. The particular solution for avoiding each of these artifacts, however, largely depends on the type of evoked potential to be measured.

As a general rule it is necessary that the patient is comfortably seated or lying down on a bed in a quiet, wll-heated, room. Adequate attention should be given to making the patient feel at ease and to explaining the entire investigation procedure.

These measures certainly contribute to lowering the "baseline" of bioelectric artifacts.

When measuring brainstem and short latency evoked potentials both spontaneous EEG and EKG activity can be largely eliminated by proper highpass filtering, e.g. at 30 Hz. Of course this is not possible when measuring longer latency (cortical) evoked potentials, for which case it is advisable to record under the eyes open condition. When measuring spinal evoked potentials it may be necessary to deliver the stimuli, triggered on the QRS complex of the EKG, in an electrically silent interval of the heart action to eliminate EKG artifact (McKay and Galloway, 1979; Dimitrijevic et al., 1978).

Eye blinks and eye movement artifact are particularly troublesome when recording long latency cortical evoked potentials. Usually, however, the potentials associated with eye movements and eye blinks are relatively large and can be dealt with through artifact rejection (cf. section 4.3). If one has sufficient recording channels available it is worthwhile, in difficult cases, to simultaneously record the electrooculogram in a different channel. This recording may then be used either for monitoring and rejecting artifact contaminated sweeps or a modified version of the oculogram recording may be used to subtract the artifact from the EEG recordings (Verleger et al., 1982; Gratton et al., 1983). In the latter case more extended computing facilities are necessary.

A basic measure to reduce artifacts due to muscle activity is to optimally relax the patient as pointed out previously. Particular attention should be paid to a proper positioning of the head, so that neck, scalp and jaw muscles are completely relaxed. It often helps to present auditory feedback of one or more recording channels to the patient. Initially the patient may be asked to exert some of the above muscles to become familiar with their characteristic sound. This may greatly help in controlling their activity during the course of the investigation.

There are several other possible artifacts, both of bioelectric and of technical origin, the most important of which are related to the electrodes and electrode connections.

Skin potentials are slowly varying potential changes which are due to a bad electrode-to-skin interface. They can be avoided by assuring a sufficiently low electrode-to-skin impedance ($<2\ k\Omega$). Also, electrodes not carefully fixed to the skin may produce large transients, commonly denoted as movement artifacts. A similar effect may be caused by intermittent breaks in electrode wires or their connections. In all these cases one should certainly not rely on a proper artifact rejection but, instead, search for and remove the cause of the artifact, even when it should be necessary to entirely replace one or more electrodes. Eventually such costs less time and effort than if one ends up with unreliable or noninterpretable recordings.

10. FURTHER GUIDELINES FOR MEASURING AND INTERPRETING EVOKED POTENTIALS

10.1 Suggested standard protocols

The purpose of the present subsection is to present a number of standard protocols, the rationale of which is described in more detail in following chapters, for obtaining various types of evoked potentials. It should be emphasized that in clinical practice protocols different from the ones described here may be more adequate, depending on the particular problem at hand. The point to make, however, is that one should aim at some standardization, particularly in view of the desirability to build up one's own normative data-set for each specific protocol (see subsection 10.3)

Table 2.2 summarizes the essential stimulus, recording and analysis parameters, as well as electrode locations. Many of the suggested parameter values may be changed within reasonable limits without basically affecting the results.

	AUDITORY EVOKED POTENTIALS			VISUAL EVOKED POTENTIALS		SOMATO SENSORY EVOKED POTENTIALS		
	brain stem	middle latency	long latency	flash	pattern reversal	short latency median nerve	short latency sural nerve	long latency
STIMULATION								
(mean) rate (Hz)	11	4	0,25	1	1	11	4	0,25
regular/random	REG	REG	RAN	RAN or REG	RAN or REG	REG	REG	RAN
duration (ms)	0,1	0,1 or 10	50 to 60	10	X	0,1	0,1	0,1
intensity	70 dB	80 dB	60 or 70 dB	X	30 cd/m^2	clear thumb twitch	supramaximal sensory AP	see short latency
notes	rarefaction clicks	clicks or tone-bursts	clicks or tonebursts		check stimuli			
RECORDING								
sensitivity (uV full scale)	10	50	100	100	100	50	50	100
high pass filter (Hz)	150	5 or 25	0,5	0,5	0,5	30	5	1
low pass filter (Hz)	3000	175 or 250	75	70	70	1500	250	250
ANALYSIS								
time base (ms)	10	50 or 100	500	500	500	50	100	500
number of sweeps	2000	100-200	64	60 or 120	60 or 120	1000	1000	100
number of channels	2	1	3	4	4	3	1	4
note: time base does not include a possible pre-stimulus interval								
ELECTRODE POSITIONS								
active ch 1	pretragus 1	M1 or M2	T3	O1	O1	Erb	Cz'	C7
active ch2	pretragus 2	-	T4	Oz	Oz	C7	-	C3'
active ch3	-	-	Cz	O2	O2	C3' or C4'	-	C4'
active ch4	-	-	-	Pz	Pz	-	-	Cz'
reference	Cz	Cz	Fz or M	Fz	Fz	A1 or A2	A1 or A2	A1 or A2
common (ground)	Fpz	M2 or M1	M2 or M1	Cz	Cz	stimulated arm	stimulated leg	see short latency
notes	For proper wave form polarity connect Cz to active input					most effective ground electrode is wrap-around fabric covered lead band		

Table 2.2 Suggested standard protocols for measuring auditory, visual and somatosensory evoked potentials.

10.2 Controlling reliability and reproducibility

As stated before, the diagnostic yield of any evoked potential measurement depends, in the first place, on optimal stimulation and recording techniques, giving artifact free, reliable and reproducible wave forms.

To assess the quality and reliability of a measured wave form in a quantitative manner, several techniques may be employed, preferably in combination with one another. The following discussion provides a brief overview of the most important ones.

I. Acquisition of Duplicate Wave forms

The acquisition of duplicate wave forms under otherwise exactly the same stimulation and recording conditions, is the minimum requirement for obtaining an impression on wave form reproducibility. It is thereby advantageous to be able to superimpose two or more wave forms for easy visual inspection. Duplicate wave forms can be obtained in a sequential manner or, alternately, by averaging even and odd responses in different channels. The latter method is instrumentally more complicated although it has the advantage that a possible trend in the evoked potential wave form will influence both averages in a similar way. A practical advantage of the former method is that on several evoked potential instruments the result of the first average can be analysed and evaluated while the second average is being acquired, thus saving time.

Although duplicate wave forms will rarely be identical, one should insist on a reasonable agreement with respect to the various components, their latencies and, somewhat less important, their amplitudes (Fig. 2.17). In appreciating possible wave form differences one should be aware of the various factors that can influence wave form reproducibility. Habituation and vigilance effects and inherent response variability are almost negligible for spinal and subcortical components. They may, to some extent, influence specific cortical components, but they generally have a large impact on nonspecific cortical components.

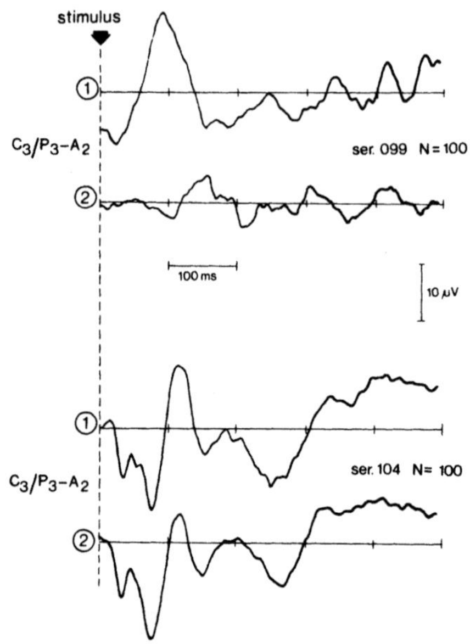

Fig. 2.17 Duplicate recordings (median nerve somatosensory evoked potentials) showing a totally irreproducible (top) and a fairly reproducible (bottom) averaged wave form.

II. Acquisition of a Dummy Average

A "dummy" average is obtained by switching off the stimulus, using an otherwise standard protocol. After a normal number of sweeps has been averaged, a virtually flat baseline should remain when the nominal gain factor is used. This procedure is particularly valuable for periodic inspection of the system. Especially technical, stimulus locked, artifacts of unexpected origin can be effectively traced in this manner. It is important that the stimulus be switched off close to the source so that as many sources of artifact as possible are included and can thus be localized. For visual stimuli, the best way is to cover the stimulating source with an untransparent dark cloth.

III. Acquisition of an alternate average

When subsequent responses are alternately added and subtracted the underlying evoked potential cancels out, so that an impression of the remaining background activity can be obtained (Fig. 2.18, see also Wong and Bickford, 1980).

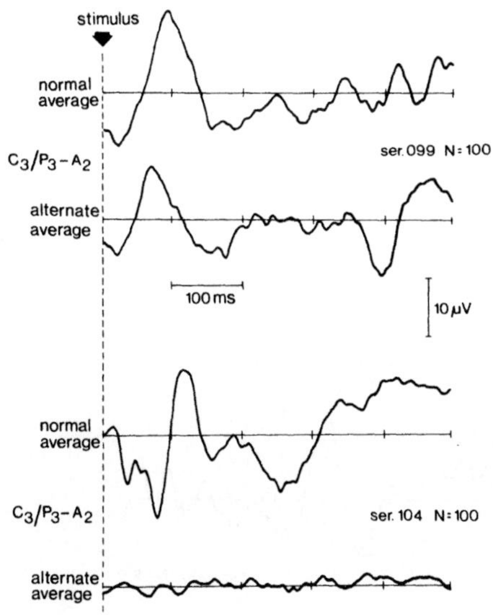

Fig. 2.18 Normal and alternate averages of the same evoked potentials as shown in Figure 2.17. Note the large wave form fluctuations in the alternate average of the upper evoked potential, indicating that the corresponding normal average is unreliable. This can be either due to large background activity or response variability.

In principle this procedure can be carried out in parallel to the normal averaging, when an additional channel is available. This allows a direct comparison of the "noise floor" thus obtained with that remaining in the normal average.

An inherent drawback, however, is that stimulus locked artifacts are not reflected in the alternate average since these are, by virtue of their consistency, cancelled out as well.

IV. Averaging a pre-stimulus interval

One of the most elegant and simple methods to obtain an impression of the background noise is to average the activity over some time prior to stimulus presentation. This method enables a fair judgement as to which part of the response is actually induced by the stimulus (Fig. 2.19).

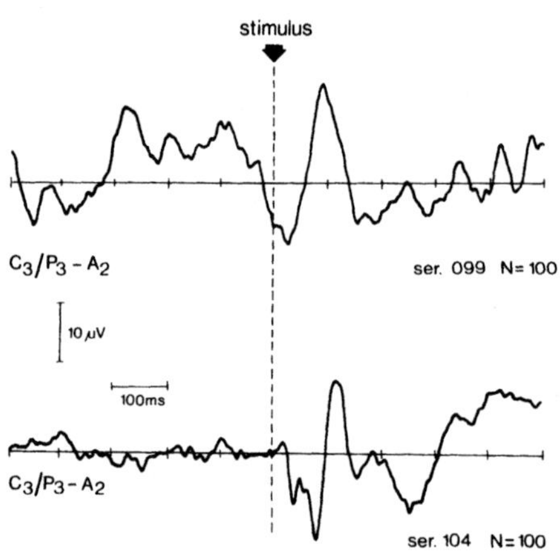

Fig. 2.19 Normal averages, including a 100% pre-stimulus interval of the same evoked potentials as shown in Figure 2.17. From the large fluctuations in the pre-stimulus interval of the upper average it must be concluded that this evoked potential is seriously contaminated by background activity.

In particular, one may obtain a qualitative impression of the reliability of various wave form components by comparing their amplitudes to the variations in amplitude occurring prior to the moment of stimulation.

In many modern evoked potential systems it is possible to include a pre-stimulus analysis interval, but if this provision is not available it can be simply technically realized by digitally delaying the stimulus (Fig. 2.20). Usual pre-stimulus intervals range from 10 to 25% of the full time base.

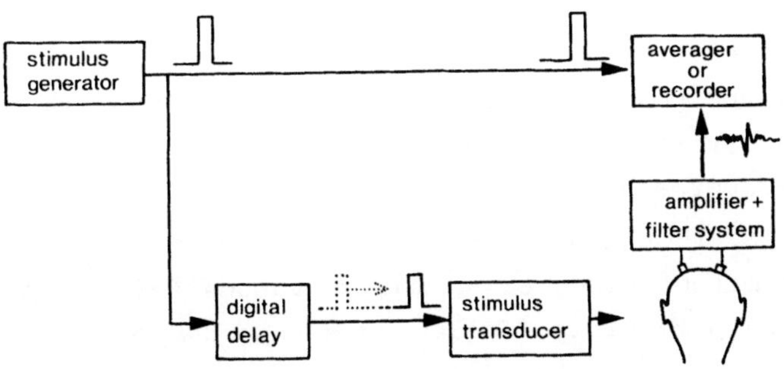

Fig. 2.20 Block diagram outlining the principle of creating a pre-stimulus interval by digitally delaying the stimulus.

10.3 General considerations on normative data and criteria for abnormality

When can an evoked potential be considered to be abnormal? In the first place one will observe wave form morphology and look for the presence or absence of various wave form components. This will provide a first indication of the degree of possible abnormality, although it should be noted that the absence of certain wave form components persé does not have to indicate pathology. Next, latencies and amplitudes (either absolute or relative) of preselected components are determined and compared to a data base of normative values. This sounds fine in theory but the actual practice certainly requires additional explanation and comment.

Comparison of actually obtained values to a set of matched normative data requires, first of all, the availability of a data base. From many scientific publications on evoked potentials one can now learn that it is a mistake to adopt normative data from other laboratories instead of acquiring one's own values (see e.g. Shearer and Dustman, 1980). There can be little doubt that this is, in essence, a wise attitude, but on the other hand the acquisition of reliable normative data of a sufficient sample size for the various evoked potential types, under different stimulation and recording conditions and as a function of age, sex and possible other factors is a life task in its own. For the average clinical laboratory this is clearly an impractical approach.

A reasonable, though not ideal, alternative procedure is to acquire a limited number of normative values for a selected set of evoked potentials, following standard protocols as described in the literature. These should then be compared to the normative values from the same literature. It is thereby advisable to include both females and males on both ends of the age range and account for a possible sex or age difference. If the sparse normative data thus obtained agree well with the normative data from literature, it is not unreasonable to adopt the latter ones *in toto*. If, on the other hand, the data do not agree well, it is worthwhile to carefully consider all the various technical and physiological factors which may have a significant impact. If the cause for deviation cannot be eliminated, the acquisition of one's own normative data is eventually needed.

Once the normative data are available, the question remains what criteria should be used for characterizing an evoked potential as either normal or abnormal. The usual way is to define some upper (or lower) boundary of normality which is derived from the underlying distribution of normative values;

more specifically, from the mean and standard deviation. This approach assumes that the parameter under consideration has a Gaussian distribution, an assumption which holds reasonably well for latency, but less so for amplitude measures. Another implicit assumption is that the mean and standard deviation derived from the normative data set represent exact, error free values. However, normative data sets are generally based on limited sample sizes (say 20-50) so that the derived mean and standard deviation are in themselves subject to variability.

It is therefore more appropriate to use so-called tolerance limits (see e.g. Stockard, et al., 1978) rather than confidence limits. Essentially, the use of tolerance limits leads to a wider range of normality since the standard deviation is scaled up by a factor that depends on the sample size of the normative group. For example, the commonly used 95% confidence limits correspond to mean ±1.96 SD, whereas the 95% tolerance limits correspond to mean ±2.38 SD (N=50), mean ±2.75 SD (N=20) or mean ±3.38 SD (N=10), where N represents the sample size of the normative data group. (For a comprehensive account on the subject, see e.g. Bowker and Lieberman, 1972, Chapter 8).

In view of the potential consequences, it might be generally stated that it is more acceptable to incur an increased number of false negatives (i.e. classification as normal while the evoked potential actually is abnormal) in favor of a reduced number of false positives (i.e. classification as abnormal, while the evoked potential is actually normal). This fact stresses the point that one should use a more conservative approach than the mean ±2 SD rule which is oftentimes used for the range of normality.

Finally, as mentioned before, latency measures tend to have significantly lower interindividual variability than amplitude measures. In fact, absolute amplitudes as such prove to be less useful in assessing evoked potential abnormality. On the other hand, intraindividual differences in amplitude (and latency) generally provide significant clinical information. It should be noted that whenever ratios (e.g. amplitude ratios) are used, the logarithm of the ratio should be taken in order to transform the distribution into Gaussian so that the above mentioned statistics become applicable.

11. GLOSSARY OF ACRONYMS

AC	alternating current, also used to indicate signals having meaningless DC component
ADC	analog-to-digital conversion
AEP	auditory evoked potential
AM	amplitude modulation, e.g. used in auditory stimulation
BAEP	brainstem auditory evoked potential
CMRR	common mode rejection ratio, used to express the attenuation of the common vs. the differential signal of an amplifier
CNAP	compound nerve action potential
dB	decibel, unit to express a relative sound intensity or a relative amplitude
DC	direct current, also used to indicate signals which have a constant amplitude over time
EEG	electroencephalography or electroencephalogram
EKG	electrocardiography or electrocardiogram
EMG	electromyography or electromyogram
EP	evoked potential
f	frequency
FFP	frequency following potential
FFT	fast fourier transform, algorithm used for transforming data from the time to the frequency domain and vice versa
FM	frequency modulation, e.g. used in auditory stimulation and in instrumentation tape recorders
Hz;kHz	Hertz, the unit for frequency; kiloHertz
HL	hearing level
IEC	International Electrical Committee, an international standardization committee
LED	Light-emitting diode, sometimes used for visual stimulatio
N_{70}	a negative wave form component with a peak around 70 ms
P_{100}	a positive wave form component with a peak around 100 ms
SL	sensory (or sensation) level
SPL	sound pressure level
s, ms, µs, ns	seconds, milli-, micro-, nano-
SNR	signal-to-noise ratio

SSEP	somatosensory evoked potential
UL	Underwriters laboratories, U.S. organisation dealing with standardization and testing of equipment
V, mV	volts, milli-, micro-, nano-
μV, nV	
VEP	visual evoked potential

REFERENCES

Allison, T.: Recovery functions of somatosensory evoked responses in man.Electroenceph. Clin. Neurophysiol., 14, 331-343 (1962).

Beagley, H.A. (ed.): Auditory Investigation, The scientific and technological Basis, Oxford Univ. Press, (1979)

Boston, J.R. and Ainslie, P.J.: Effects of analog and digital filtering on brainstem auditory evoked potentials. Electroenceph. Clin. Neurophysiol. 48, 361-364 (1980).

Bowker, A.H. and Lieberman, G.J.: Engineering Statistics, Prentice Hall, Englewood Cliffs, (1972).

De Boer, E.:Quantisation and averaging - a crude but suggestive analysis.Scand. Audiol., Suppl. 9, 109-110 (1979).

Desmedt, J.E., Brunko, E., Debecker, J. and Carmeliet, J.: The system bandpass required to avoid distortion of early components when averaging somatosensory evoked potentials. Electroenceph. Clin. Neurophysiol., 37, 407-410 (1974).

De Weerd, J.P.C. and Martens, W.L.J.: Theory and practice of a posteriori "Wiener" filtering of average evoked potentials. Biol. Cybern., 30, 81-94 (1978).

De Weerd, J.P.C.: Facts and fancies about a posteriori "Wiener" filtering.IEEE Trans. Biomed. Engng., BME-28, 252-257,(1981a).

De Weerd, J.P.C.: A posteriori time-varying filtering of averaged evoked potentials.I. Introduction and conceptual basis. Biol. Cybern., 41, 211-222 (1981b).

Dimitrijevic, M.R., Larsson, L.E., Lehmkuhl, D. and Scherwood, A: Evoked spinal cord and nerve root potentials in humans using a noninvasive recording technique. Electroenceph. Clin. Neurophysiol., 45, 331-340 (1978).

Doyle,D.J. and Hyde, M.L.: Analogue and digital filtering of auditory brainstem reponses. Scand. Audiol., 10, 81-89 (1981a)

Doyle, D.J. and Hyde, M.L.: Bessel filtering of brainstem auditory evoked potentials. Electroenceph. Clin. Neurophysiol, 51, 446-448 (1981b).

Fridman, J., John, E.R., Bergelson, M., Kaiser, J.B. and Baird, H.W.: Application of digital filtering and automatic peak detection to brainstem auditory evoked potential. Electroenceph. Clin. Neurophysiol., 53, 405-416 (1982).

Gabriel, S., Durrant, J.D., Dickter, A.E. and Kephart, J.E.: Computer identification of waves in the auditory brainstem evoked potentials. Electroenceph. Clin. Neurophysiol., 49, 421-423, (1980).

Geddes, L.S.: Bioelectrodes. Am. J.EEG Technol., 15, 99-106, (1975).

Gratton, G., Coles, M.G.H. and Donchin, E.: A new method for off-line removal of ocular artifact. Electroenceph. Clin. Neurophysiol., 55, 468-484, (1983).

Handbook of Clinical Neurophysiology, Volumes 3, 4 and 8. Remond, A. (Ed.-in-chief): Vol 3: Techniques and Methods of Data Acquisition of EEG and EMG, DeLucchi, M.R. (Ed.), Vol 4: Evaluation of Bioelectrical Data from Brain, Nerve and Muscle, I. Brazier, M.A.B. (Ed.), Vol 8: Electrical Reactions of the Brain and Complementary Methods of Evaluation. Buser, P. (Ed.), Elseviers Sc. Publ. Cy., Amsterdam. The Netherlands, (1978).

International Electrical Committee IEC 601-1: Safety of medical electrical equipment, Part 1: General requirements. IEC, Geneva, (1977).

Jackson, S.A. and Barber, C.: The effect of temporal stimulus parameters upon the VEP. In: Evoked Potentials, Proceedings of Int. Evoked Potentials Symposium, Nottingham, MTP Press, Lancaster, (1980).

Jewett, D.L. and Williston, J.S.: Auditory evoked far fields averaged from the scalp of humans. Brain, 94, 681-696, (1971).

McGill, K.C., Cummins, K.L., Dorfman, L.J., Berlizot, B.B., Luetkemeyer, K., Nishimura, D.G. and Widrow, B.: On the nature and elimination of stimulus artifact in nerve signals evoked and recorded using surface electrodes. IEEE Trans. Biomed. Engng., BME-29, 129-137, (1982).

McKay, W.B. and Galloway, B.L.: Technological aspects of recording evoked potentials from the cauda equina and lumbosacral spinal cord in man. Am. J. EEG Technol., 19, 83-96, (1979).

Michelini, S., Arslan, E., Prosser, S. and Pedrielli, F.: Logarithmic display of auditory evoked potentials. J. Biomed. Engng., 4, 62-64, (1982).

Møller, A.R.: Improving brainstem auditory evoked potential recordings by digital filtering. Ear and Hearing, 4, 108-113, (1983).

National Fire Protection Association: Electricity in patient care facilities. NFPA No. 76 B-T, Boston, (1973).

Pratt, H., Politoske, D. and Starr, A.: Mechanically and electrically evoked somatosensory potentials in humans: Effects of stimulus presentation rate. Electroenceph. Clin. Neurophysiol., 49, 240-249, (1980).

Pratt, H., Ben-David, Y., Peled, R., Podoshin, L. and Scharf, B.: Auditory brainstem evoked potentials: Clinical promise of increasing stimulus rate. Electroenceph. Clin. Neurophysiol. 51, 80-90, (1981).

Regan, D.: Fourier analysis of evoked potentials: Some methods based on Fourier analysis. In: Visual evoked potentials in man: New developments. J.E. Desmedt (Ed.), p. 110-117 , Clarendon Press, Oxford, (1977).

Rossini, P.M., Cracco, R.Q., Cracco, J.B. and House, W.J.: Short latency somatosensory evoked potentials to peroneal nerve stimulation: Scalp topography and the effect of different frequency filters. Electroenceph. Clin. Neurophysiol., 52, 540-552, (1981).

Rothman, H.H., Davis, H. and Hay, I.S.: Slow evoked cortical potentials and temporal features of stimulation. Electroenceph. Clin. Neurophysiol., 29, 225-232, (1970).

Ruchkin, D.S.: An analysis of average response computations based upon aperiodic stimuli. IEEE Trans. Biomed. Engng., 12, 87-94, (1965).

Salamy, A., McKean, C.M., Pettett, G. and Mendelson, T.: Auditory brainstem recovery processes from birth to adulthood. Psychophysiology, 15, 214-219, (1978).

Scherg, M.: Distortion of the middle latency auditory response produced by analog filtering. Scand. Audiol., 11, 57-60, (1982).

Shearer, D.E. and Dustman, R.E.: The pattern reversal evoked potential: The need for laboratory norms. Am. J.EEG Technol., 20, 185-200, (1980).

Spekreyse, H., Estevez, O. and Reits, D.: Visual evoked potentials and the physiological analysis of visual processes in man. In: Visual evoked potentials in man: New developments J.E. Desmedt (Ed.), p. 16-89, Clarendon Press, Oxford, (1977).

Stapells, D.R. and Picton, T.W.: Technical aspects of brainstem evoked potential audiometry using tones. Ear and Hearing, 2, 20-29, (1981).

Stockard, J.J., Stockard, J.E. and Sharbrough, F.W.: Non-pathologic factors influencing brainstem auditory evoked potentials. Am. J. EEG Technol., 18, 177-209 (1978).

Stockard, J.J., Hughes, J.F. and Sharbrough, F.W.: Visually evoked potentials to electronic pattern reversal: Latency variations with gender, age and technical factors. Am.J. EEG Technol., 19, 171-204, (1979).

Stockard, J.E. and Westmoreland, B.F.: Technical considerations in the recording and interpretation of the brainstem auditory evoked potential for neonatal neurologic diagnosis. Am. J. EEG Technol., 21, 31-54, (1981).

Underwriters Laboratories: UL544: Standard for medical and dental equipment. New York, (1980).

Van Olphen, A.F., Rodenburg, M., Verwey, C.: Influence of the stimulus repetition rate on brainstem evoked responses in man. Audiology, 18, 388-394, (1979).

Verleger, R., Gasser, T. and Möcks, J.: Correction of EOG artifacts in ERP of the EEG: Aspects of reliability and validity. Psychophysiology, 19, 472-480, (1982).

Walter, D.O.: A posteriori "Wiener filtering" of average evoked responses. Electroenceph. Clin. Neurophysiol., Suppl. 27, 61-70, (1969).

Wastell, D.G.: The application of low-pass linear filters to evoked potential data: Filtering without phase distortion. Electroenceph. Clin. Neurophysiol., 46, 355-356, (1979).

Wastell, D.G.: Temporal uncertainty and the recovery function of the auditory EP. In: Evoked Potentials, Proceedings of Int. Evoked Potentials Symposium, Nottingham, MTP Press, Lancaster, (1980).

Wong, P.K.H. and Bickford, R.G.: Brainstem auditory evoked potentials: The use of noise estimate. Electroenceph. Clin. Neurophysiol., 50, 25-34, (1980).

Woods, D.L., Courchesne, E., Hillyard, S.A. and Galambos, R.: Recovery cycles of event-related potentials in multiple detection tasks. Electroenceph. Clin. Neurophysiol., 50, 335-347, (1980).

Yagi, T. and Kaga, K.: The effect of the click repetition rate on the latency of the auditory evoked brainstem response and its clinical use for a neurological diagnosis. Arch. Otorhinolaryngol., 222, 91-97, (1979).

Chapter III

Auditory Evoked Potentials

A. Zonneveldt
J.L. Blom
J.J. Rotteveel

1. INTRODUCTION

The Auditory Evoked Potential (AEP) has a long history in human neurofysiology. With high intensities a signal can sometimes be seen in temporal leads in the EEG. It is questionable however, if the response is generated by the brain. More probable it is the myogenic manifestation of a reflex which is elicited with high intensity sounds. One of the first reports in literature is from Davis (1939) who investigated the effect of acoustic stimuli on the EEG. Although some effect was discernable, its main effect was desynchronisation of the traces without specific activity attributable to the sound stimuli. It was not until the advent of averaging techniques that a more general interest in the AEP arose. Its first use was in the field of audiology with great expectations for objective audiometry. These were soon tempered by the realisation that the technique to acquire reliable results in Evoked Response Audiometry (ERA) was very difficult and also sufficient knowledge was lacking for interpretation of the obtained signals. The many different laboratories, which in the end of the fifties and begin of the sixties applied the new technical possibilities of special purpose computers to the averaging of brain signals, were often disappointed because of the seemingly conflicting results which made the usefulness of ERA doubtful.

In the mean time other techniques have evolved rendering cortical ERA less generally applicable. To assess the periferal part of the auditory system cochleography can be used, although its invasive nature, necessitating sedation, is a disadvantage. To acquire more insight in the events occurring in the primary processing of the auditory system in the subcortical areas, the Brain Stem Auditory Evoked Potential (BAEP) has quickly

become one of the most important tools for clinical use.
The cortical AEP has found other applications which are somewhat outside the field of clinical neurology. In psychology and psychiatry it has useful applications; one example is the study of attention in which often the AEP is used, another is its use in repeated studies of psychiatric patients.

For clinical neurophysiology its use is limited. When diagnostic problems arise in the auditory system or in the temporal lobe, the AEP may be very useful. For the assessment of more general disturbances of the CNS the VEP is better known and therefore the first choice, although stimulus parameters are much easier controlled for the AEP than for the VEP.

As has been mentioned the best known auditory evoked response is the brain stem response, first described by Jewett et al. (1971) and the cortical AEP. A 'middle latency response', covering the period between 12 and 50 milliseconds after the stimulus, has been described by several authors and some think it the best candidate to be used in evoked response audiometry. Up till now, however, it is not used in clinical neurofysiology and therefore little attention will be given to this response.

2. ANATOMY AND PHYSIOLOGY

2.1 The transducer

Sound is represented as pressure waves in air which have to be converted to action potentials. The transducing system consists of the middle ear and the cochlea. The first is a mechanical amplifier composed of the eardrum, the three ossicles, hammer (*malleus*), anvil (*incus*) and stirrup (*stapes*), which transmit the tympanic movements to the membrane of the oval window. Its vibrations excite pressure waves in the perilymfatic fluid within the membraneous cochlea. The latter is a coiled tube, divided in three compartiments, the *scala vestibuli*, upper part and *scala tympani*, lower part, connected at the top and filled with perilymphe, and the *scala medium*, filled with endolymphe (see Fig. 3.1).

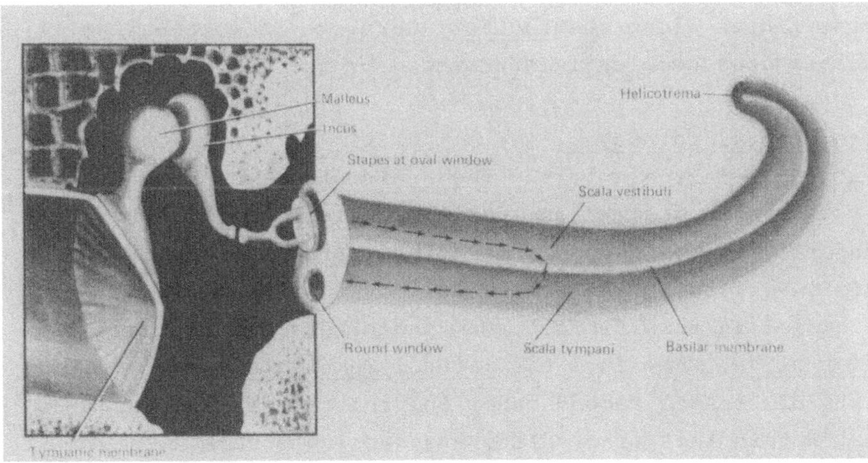

Fig. 3.1 Schematic view of the auditory sense organ. The tympanic membrane adheres to the malleus, the first of the three ossicles which form the connection to the oval window. Movements of the eardrum are transferred through this osseous chain which acts as a mechanical impedance adaptation to the fluid filled cochlea. Inward movements of the oval window sets up pressure waves which travel the whole length of the cochlea, exciting those receptors which resonate with the frequency of excitation. Damping of the waves is due to the flexible round window which bulges into the middle ear cavity. (reproduced from 'Principles of Neural Science' Kandel&Schwartz, Elsevier, 1981, with permission).

The basilar membrane separates the scala media and scala tympani and bears the actual transducers, Corti's organs. They are neatly ordered along the whole tube. In the basilar membrane ± 20.000 stiff elastic fibres project transversely from the centre spill, the modiolus, to the outer wall, their lenght increasing towards the top of the cochlea. The pressure waves in the perilymphe let the appropriate basilar fibres resonate, thereby moving the hairs of the hair cells against the tectorial membrane. This produces potential changes which as receptor potentials inducing action potentials in the nerve endings surrounding the bases of the hair cells. The cell bodies of these axons form the spiral ganglion, sending their efferent fibres along the auditory nerve to the brain stem. The membraneous cochlea is surrounded by the bony cochlea.

2.2 The central pathways.

All acoustic fibres entering the brain stem synaps on two cell masses, the ventral and dorsal cochlear nucleï (Fig. 3.2). The majority of the second order fibres from the nuclei cross the midline, forming the trapezoïd body, where some synaps on small nucleï in this fibre tract and others end in the superior olive. They form the lateral lemniscus which mainly consists of crossed second order and third order auditory fibres. Within this tract diffuse synapsing occurs on neuronal aggregates. This compound bundle ends at the inferior colliculus, a small number of fibres passing uninterrupted to end in the medial geniculate body. A thick fibre bundle, the brachium of the inferior colliculus, connects the latter to the diencephalic auditory relay station, the medial geniculate body.

The connection to the cortex is formed by the auditory radiation, which projects to the primary auditory area found in the transverse temporal gyrus of Henschl (area 41 of Brodman). This centrifugal fibresystem is parallelled by a centrifugal system from which the final efferent part is formed by the olivo-cochlear bundle ending directly on Corti's organs.

An important aspect of the auditory system is its tonotopic localisation. This denotes the sequential arrangement of orderly increasing frequencies which is not only found in the cochlea but in all nucleï in the auditory system.

2.3 Physiology.

The human ear is responsive to pressure waves with frequencies between 18 and 21000 Hz. This upper limit decreases after adulthood to ± 15000 Hz at 35 year and below 6000 Hz at ages over 70. The most important frequencies are centered around 1000 Hz (100-8000 Hz) containing the sound spectrum of human speech.

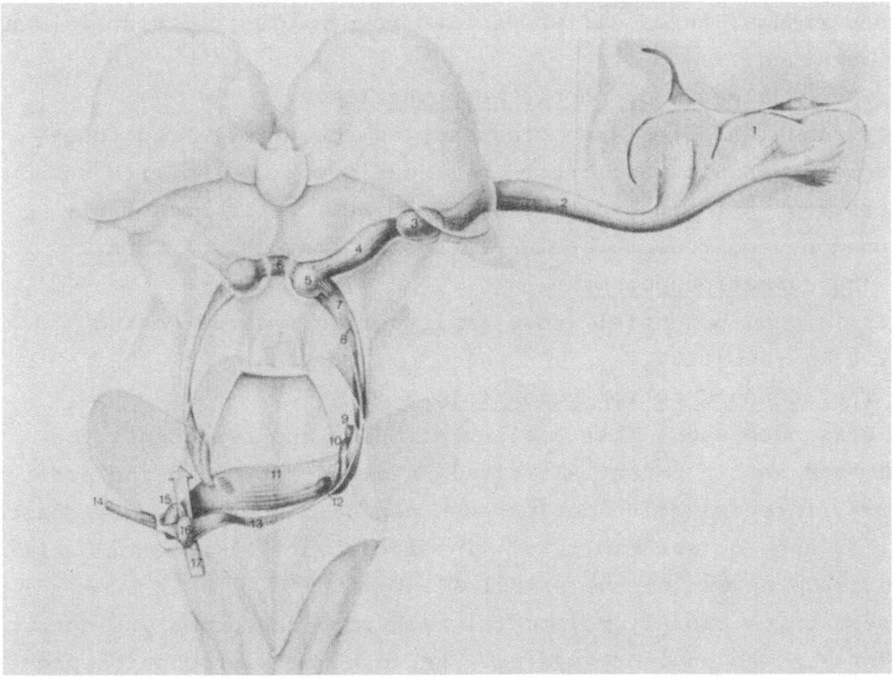

Fig. 3.2 Schematic view of the auditory system projected on the brainstem.

1. gyri temporalis transversales
2. radiatio accoustica
3. corpus geniculatum mediale
4. brachium colliculi inferius
5. colliculus inferior
6. commissura colliculus inferior
7. lemniscus lateralis
8. nucleï lemniscus lateralis
9. nucleus lateralis oliva superior
10. nucleus medialis oliva superior
11. corpus trapezoïdeum
12. nucleï corpus trapezoïdeum
13. striae acusticae dorsales
14. nervus cochlearis
15. nucleus cochlearis ventralis
16. nucleus cochlearis dorsalis
17. pedunculus cerebellaris inferior

(Reproduced from 'The Human Central Nervous System', Nieuwenhuys et al. 1979, Springer, with permission).

When sound is presented to the ear with an electrode at the round window, three different electrophysiological signals can be recorded.

1. The cochlear microphonic response (CM)

This signal has the same frequency as the stimulating tone-frequency and its amplitudo is linearly correlated with sound pressures up to intensities of 80 dB SPL. It is presumably caused by the receptor potentials of the organs of Corti.

2. The summation potential

This is a DC-potential whose amplitudo is related to the stimulus intensity.

3. The compound action potential

This is best seen after a click stimulus and represents the summated action potentials of all excited fibres in the auditory nerve. Three negative deflections can be discovered, designated N1, N2 and N3 (see Fig. 3.3). The latency of N1 is about 1 ms and is dependent on the stimulus intensity.

These three closely related signals form the primary electric events in the auditory system. The cochlear microphonics presumably are the source for the action potentials generated in the nerves of the spiral ganglion which act like other neurons. Frequency content of the signal is carefully encoded in the impulse sequences occurring in groups of fibres. Individual fibres exhibit characteristic frequencies which are those frequencies for which the lowest threshold exists. Tonotopy relates to these characteristic frequencies, the highest originating in fibres connected to receptors at the beginning of the cochlea near the oval window, the lowest being found at the apex. These signals can be recorded 1-2 ms after the stimulus in the cochlear nucleï. Here the first signal processing takes place intended to increase frequency discrimination.

From this station signals spread in the lower brain stem, via a number of parallel paths. One way projects bilaterally to the superior olives and presents the first link for the timing sequences necessary for directional hearing. The next station is formed by the trapezoïd body and superior olive where second order signal processing takes place; there the mean frequency is much lower than in the cochlear nuclei. There again, signal-processing leads to an increase in discriminative power.

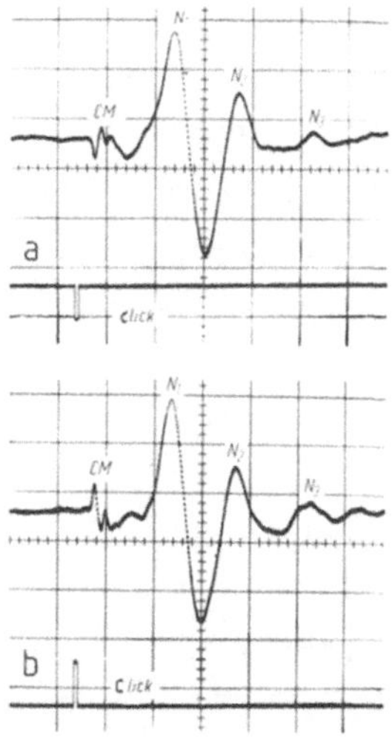

Fig. 3.3 Cochlear potentials recorded from the round window. CM is the cochlear microphonic potential. It can be seen that, depending on the polarity of the click - and the movement of the earphone membrane and thus the movement of the eardrum - the CM starts either with a positive (a) or negative (b) polarity. It can also be seen that an increase in intensity is followed by an increase in CM; the nerve potentials N1, N2 and N3 are barely influenced by this change in intensity.

The superior olive is also an important control centre for sound localisation. This can be seen, among other things, in its widespread connections with efferent motor nucleï (trigeminal and facial) which innervate the ear muscles. As the lemniscal nucleï are situated higher up in the brainstem along the lateral lemniscus their responses occur later, giving rise to complex time delays in the auditory path to the inferior colliculus where practically all signals have their third or fourth level of signal processing. Here sharply tuned as well as broad-band neurons can be found, the latter being especially sensitive to frequency changes. The mean frequency of most neurons show an approximately linear relation to intensity, while also an increasing number of neurons is excited with increasing intensity. As in the superior olive,bilateral innervation can also be found at this,level and many intricate mechanisms show that especially direction of hearing is a quality with which the brain stem system can amply cope. The signals go from here to the medial geniculate body. Its precise function is only partly known as yet. It is clear that a rather extensive signal processing takes place related to directional hearing, frequency discrimination and intensity differentiation. As in the other systems at the diencephalic level, 'programmed' activity, lasting several milliseconds, is found. This programmed activity is still more pronounced at the cortical level. Here activity can be measured ± 10 ms after a stimulus is delivered to the ear and constitutes the primary part of the long latency AEP. The ensuing components found in the AEP are the result of feedback signals between the cortex and lower signal processing areas, e.g. the reticular formation. The neurons in the primary receptor areas show different kinds of reaction to sound stimulation. Detailed information must be found elsewhere. (M.Monnie Functions of the Nervous System, vol 3, Sensory Functions and perception. Elsevier, Amsterdam, London, New York, 1975, pp.1040)

3. METHOD OF STIMULATION

Stimulus parameters have to be defined in terms of time length, intensity and frequency. Intensity can be given in absolute or relative units. Sound pressure can be expressed in Pascal (Pa) in the SI-system (or dyn. cm^{-2}) and is a measure of the force exerted on the eardrum by the stimulus. It is, however, rather unpractical as it is quadratic and has a range of more than 10^7. A more practical way is to express sound intensity as the ratio of two sound pressure values. As the latter are quadratic their logarithm transform is linear. This measure is the Bell, defined as the decadic logarithm of the quotient of two sound pressure values. The decibel (dB) is 0.1 Bell.

In audiology different intensity scales are used, all based on different reference values for the ratio. The most absolute dB scale is the dBA-scale. There the reference value is an internationally defined reference value of 2×10^{-5} Pa (2×10^{-4} dyn cm^{-2}). There A stands for absolute. It is also given as dB SPL (sound pressure level).

The human auditory system is not equally sensitive to all frequencies. It is most sensitive between 2 and 4 Kc, therefore when dBA is used it is necessary to indicate the frequency.
A different scale is introduced for loudness, the Phone, which indicates a loudness equal to the dBA-value of a tone of 1000 Hz. A 2000 Hz tone, for example, with an intensity of 50 Phone has a loudness equal to a 50 dBA tone of 2000 Hz although its intensity is 10 dB higher.

In clinical neurophysiology, the hearing level scale (HL) is very often used. Here the intensity of the stimulus is referred to the mean hearing threshold of normal subjects. Also the sensation level (SL) is used, which is the sound intensity referred to the subjects own hearing threshold. Therefore the hearing threshold for the stimulus is determined with an audiometer. The stimulus intensity is then, for example, 50 dB above threshold value. This scale has the advantage that the cerebral activity between subjects can be compared as the individual loudness level for each subject is equal.

3.1 Click-stimuli

The click-stimulus is a short square-shaped impulse delivered to an earphone or loudspeaker. Its intensity must be measured related to the soundpressure it releases, which can reach the eardrum effectively. It is not sufficient to measure only the peak current or total current delivered to the output device as the amount of electrical energy converted to acoustical energy varies considerably from instrument to instrument. The sound pressure should be measured with a standard dB-meter at a standard distance to the sound source. For the BAEP usually a standard intensity of 70 dB HL is used. When hearing difficulties are met it is better first to determine the individual threshold and to deliver a stimulus relative to this threshold at 70 dB SL. The time duration of the click-stimulus must be less than 1 millisecond and is mostly between 100 and 500 microsec.

The interstimulus interval is another important parameter. Repetition rates differ considerably for different applications. Analyzing the BAEP, it is only the first 10 ms after the stimulus that are relevant, and repetition rates up to 30 per second are normally used. When, however, cortical responses are sought for, repetition rate must be much lower. An average of 2 seconds between stimuli is necessary to obtain reliable responses. Also important is the use of random interstimulus intervals to prevent the occurrence of time-lock to the background EEG-activity. Especially for the BAEP it is important to use stimuli which move the earphone or loudspeaker membrane only in one direction. When the membrane is moved towards the eardrum these stimuli are called condensation clicks, when moved the other way, rarefaction clicks. The BAEP's to either stimuli differ. The number of stimuli differs also for BAEP and clinical AEP. The former needs at least 1000 to 2000 stimuli to obtain reliable results, whereas for the cortical AEP 100 or less stimuli are sufficient. The use of advanced signal analysis techniques like time-varying Wiener filtering and Kalman filtering probably can reduce the number of stimuli needed.

3.2 Burst stimuli

Both tone bursts and noise bursts are used as stimuli to evoke an auditory response. Both require that onset and offset are gradual. When the stimulus is abruptly switched on or off switching transitions equal to the click-responses will mask effects due to different tone frequencies. With short lasting stimuli frequency-dependent tone bursts are not effective without special precautions. The solution is to use ramps of gradually increasing intensity with the on- and offset of the stimulus. For short-lasting stimuli a compromise has to be found. Various ramp-times are given in the literature. In our laboratory we use ramp-times of 5 or 15 ms.

Noise burst stimuli are not used very much for normal clinical investigations. Tone bursts, on the other hand, are the kind of stimuli that are mainly used for cortical evoked responses.

In most laboratories pure sine waves with ramps between 5 and 25 ms are used and total durations of between 15 and 150 ms. Depending on the research goal, amplitudes vary between threshold intensity and 80 to 90 dB HL or SL.

For routine clinical use, tone frequencies of 1000 Hz are used in the assessment of neurological dysfunctions in the cortical AEP. Its intensity is between 60 and 70 dB HL and its duration about 100 ms with rise and fall times of 10 ms. The number of stimuli is 100 and the interstimulus interval varies between 1.5 and 2.5 at random. The stimulus is presented to both ears simultaniously. In the response, asymmetries in latencies and large amplitude differences may have clinical significance.

For cortical Evoked Response Audiometry (ERA) mostly 64 or 100 stimuli are used with durations between 100 and 500 ms and interstimulus intervals of 1.5 and 2.5 seconds at random. The most effective application is to analyse the ERA on-line at an intensity of 30 dB and to compare the latencies and amplitudes with standard values from literature. It is common practice, before using cortical ERA clinically, to assess one's own standard values and to compare these with literature. In paragraph VII standard values will be presented. Threshold values can than be determined with a second or third recording.

3.3 Amplitude modulated stimuli

Amplitude modulated stimuli can be seen as a special form of burst stimuli. It is a continuous tone of e.g. 1000 Hz which is amplitude modulated with a low frequency sine wave mostly between 2 and 20 Hz. Because of its form no fast transitions will occur. They are the acoustic pendant of the sinesoid modulated light stimulus which is used in research on the visual system. The difference in intensities at the positive and negative crests of the modulating sinus is the stimulus intensity.

When the modulation is 100% (at the negative peak of the sinewave the amplitude is zero), the stimulus bears much resemblance to a tone burst stimulus. Amplitude modulated stimuli are seldom used in clinical practice. Their significance lies in the research of the auditory system.

3.4 Frequency modulated stimuli

The kinds of stimuli described above elicit responses due to the amplitude discriminative properties of the auditory system. These are mainly Corti's organ properties which are preserved and even enhanced in the rest of the auditory system. Another property is frequency discrimination, which is much less determined by the properties of the transducers (Corti's organ) which for low frequencies exhibit low discriminative powers. The subsequent processing stages improve this aspect considerably. From animal experiments it has become clear that the cortex and the medial geniculate body are most important. Frequency modulat tones therefore give information on different properties of the auditory system. They are elicited by shifting a tone with a continuous amplitude to a lower or higher frequency and back. The same precautions as for tone bursts have to be taken. Therefore to prevent click stimulation the change in frequency has to be gradual, and not abrupt, using ramps with a duration between 10 and 500 ms. Also the interstimulus interval time has to be sufficiently long i.e. longer than 1.5 sec and the stimuli must come irregular to prevent habituation. In Fig. 3.4 the form of the stimulus is indicated.

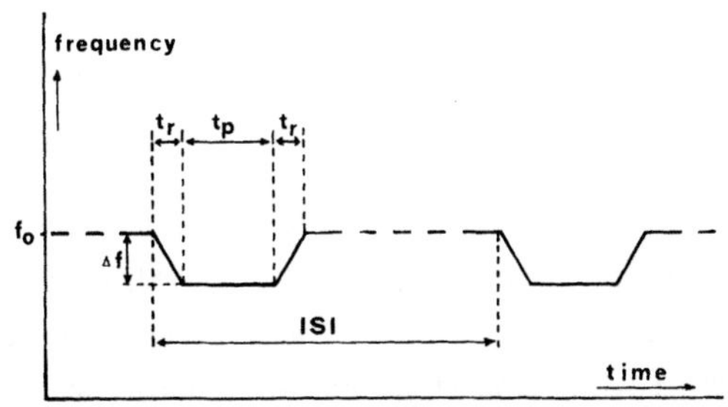

Fig. 3.4 Timing diagram of the FM modulated stimulus. t_r= ramp-time.At the beginning of t_r the frequency changes continuously to the start of t_p = the plateau time. This is the duration the tone stays on the lower frequency. At the beginning of the second t_r the tone changes frequency again to its original frequency. ISI is the interstimulus interval time. During the whole period of stimulation the tone remains audible in the same intensity. In routine applications frequency changes of 10%, ramp-times of 5 ms and plateau times of 10 ms are customary.

This type of stimulus is not yet much in use clinically although it can be advantageous in patients with centrally determined hearing disturbances. Some authors have shown its use in cortical ERA (Funasake 1974). An impediment to its use is the special stimulator, which is not yet commercially available.

4. METHOD OF ANALYSIS

4.1 BAEP

4.1.1 Electrode positions.

In recording the BAEP ordinary silver/silver chloride EEG electrodes are used,which are attached to the scalp by means of collodium, the electric contact being made by electrode jelly, after which the interelectrode impedance must be less than 3 kilo ohm. Registration is performed simultaneously from vertex to left and right ear. The term Cz-Aipsi stands for derivation from the vertex to the stimulated ear (Aipsi = ipsilateral) and Cz-Acontro for derivation to the non-stimulated ear (Acontro = contralateral). A vertex positive signal is recorded as an upward deflection (see Fig. 3.5).

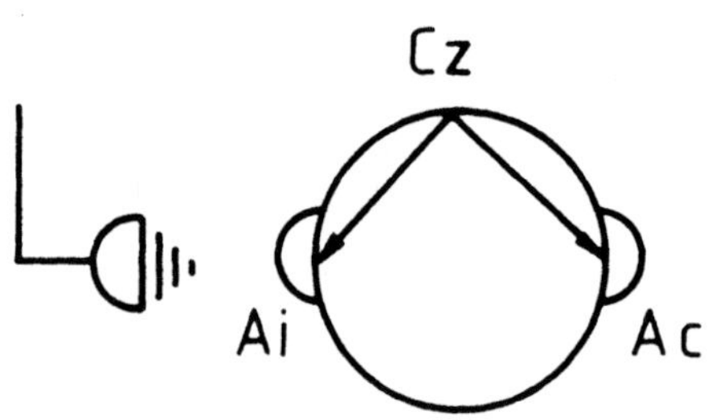

Fig. 3.5 Electrode placement.

- vertex electrode. It is generally assumed that the maximum of the BAEP is registered on the vertex with the potentials getting lower in the direction of nasion, inion and mastoid. (van Olphen et al. 1977).
- ear electrode. Wave I is believed to be caused by the nerve action potential of the acoustic nerve and therefore to be a near field potential. If this is true, the position of the electrode is critical.

To test this we recorded simultaneously from Cz to the following electrode positions:

a. the mastoid.
b. the back of the earlobe.
c. the front of the earlobe.
d. the tragus (just in front of the meatus acusticus externus).

(see Fig. 3.6).

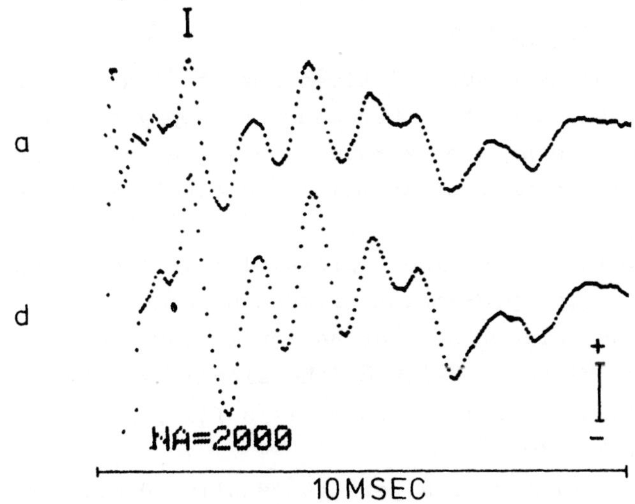

Fig. 3.6 a. electrode on the mastoid.
b. electrode on the tragus.

The highest amplitude was registered on the tragus, the lowest on the mastoid and on the back of the earlobe. We therefore advise placement of the electrode always on or near the tragus. (see Fig. 3.7).

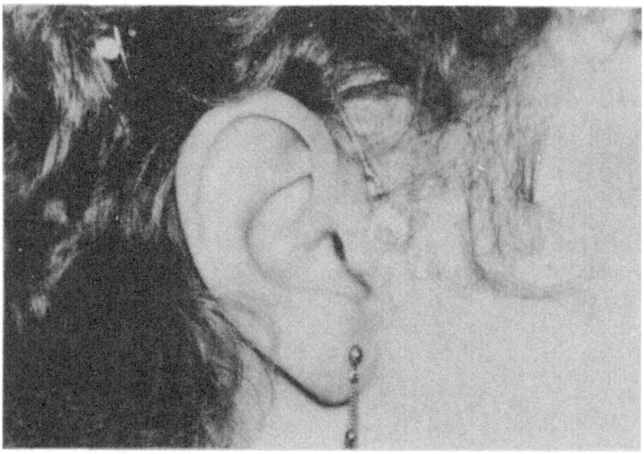

Fig. 3.7 Placement of the electrode on tragus.

4.1.2 Instrumentation

The BAEP signal is a hundred times smaller than the EEG signal. In recording the BAEP it is necessary to apply an averaging technique. This technique improves the signal/noise ratio with a factor $\sqrt{n}$ (n= number of stimuli) and consists of the following elements:

- amplifiers. These comprise a pre-amplifier and a main amplifier (differential). The pre-amplifier must be placed as closely as possible to the patient and has an amplifying factor of 10^4.
- filters. The highest energy of the signal is between 800-1000 Hz. A bandpass of 150-3000 Hz is sufficient to obtain a reliable registration.
- averager. The time of analysis is usually 10 milliseconds; the number of sweeps usually amounts to 2000.
- stimulator. The two important functions of the stimulator are the adjustment of the stimulus parameters and the generation of the stimulus. To elicit the BAEP a rarefaction click with

a duration of 100 microseconds is used. Frequency and intensity depend on the clinical indication. The impulse is given by a headphone. Monaural stimulation is preferable (one side at a time) with the other ear masked by a white noise.

- display. A hard copy of the curves is necessary. X-Y-recorders are employed in registering the curves. Sometimes polaroid photos or ultra violet recorders are used.
- For more detailed technical information see chapter 2 of this book.

4.1.3 Stimulus parameters (clicks)

A click is generated by supplying a square pulse to a headphone. The following stimulus parameters influence the form of the signal to various degrees:

- duration of the stimulus. This has no great influence as long as it is less then 1 ms; it is set in most cases on 100 microseconds.
- stimulus polarity. The best differentiated curves are found with stimulation by means of rarefaction pulses. Some laboratories, however, use condensation pulses. The normal values must then be adjusted. The use of alternating pulses may be used as a last ressort in the case of a very large stimulus artifact. However, in this case it is not possible to apply the usual normal values.
- stimulus rate. The rate of the stimulus has a distinct influence on peak latency. When the rate is increased latency increases as well. Stimulus rates in the range of 11-31 Hz in neurologic and 31 Hz in audiologic examinations are often used.
- stimulus intensity. This is the parameter with the largest influence on the BAEP. With an increase in stimulus intensity the amplitudes increase, whereas latencies decrease. This goes to a lesser degree for an intensity over 50-60 dB above hearing threshold. (Stockard et al. 1978). (see Fig. 3.8).

The stimulus intensity is usually determined with regard to the average hearing level in young adults (HL = hearing level). It also can be done with respect to the individual hearing level of the patient (SL = sensation level) or, when expressed as a physical value, sound pressure level (SPL). In neurologic examination it is usually 70 dB HL, in estimation of the hearing acuity it is 30 dB HL. (see 3 and 3.1).

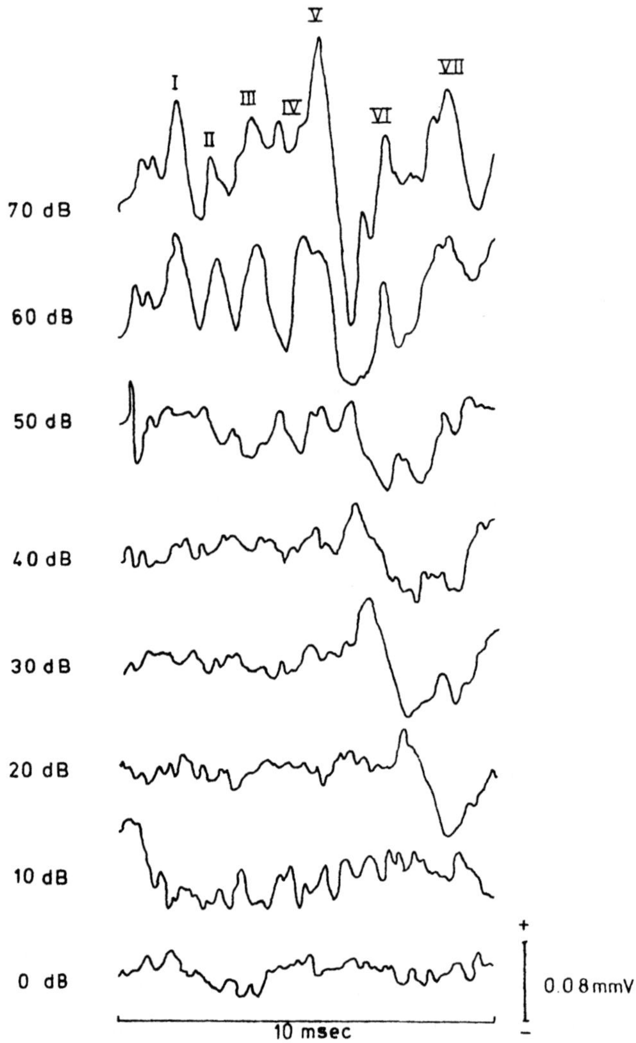

Fig. 3.8 The influence of the stimulus intensity.

4.1.4 Recording technique.

Introduction.

The patient should relax as much as possible while the examination is performed in a sound-proof room with a well-balanced temperature. Monaural stimulation is applied with the other ear masked by a white noise of 40 dB HL.

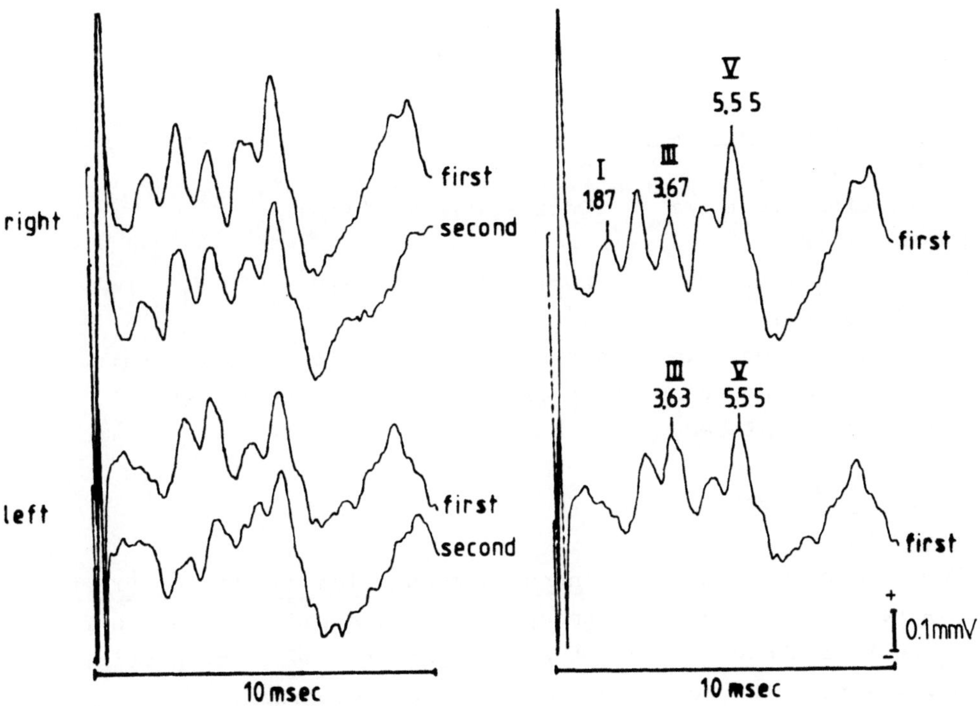

Fig. 3.9 Routine registration of a BAEP.
- female 75y right monaural stimulation 70 dB HL.

Recommendations for a test set-up in neurologic examination, to determine peak and interpeak latencies.

Stimulation right (70 dB HL, 11 Hz) → Derivation Cz - Aipsi
→ Derivation Cz - Acontro

Stimulation left (70 dB HL, 11 Hz) → Derivation Cz - Acontro
→ Derivation Cz - Aipsi

In order to judge reproducibility, the signal has to be recorded twice, both in stimulation on the right and on the left side. (see Fig. 3.9).

Estimation of the auditory acuity, to determine the hearing threshold.

Stimulation right ear (30 dB, 31 Hz)	Derivation Cz - Aipsi
Stimulation left ear (30 dB, 31 Hz)	Derivation Cz - Aipsi

When the peak of wave V is not clearly visible, the intensity should be increased in steps of 5 or 10 dB until a distinct wave V is registered.

4.1.5 Peak detection.

Latencies.

The time between the beginning of a stimulus and the occurrence of a peak is called the latency. It is measured in relation to the peak of the wave and not with regard to the first distinct deflection of the zero line (as in electromyography).

The latencies of all waves I to VII are thus measurable. (see Fig. 3.11A). Sometimes a peak is rather flat or deformed, which makes it hard to distinguish the point of inflection (top). In that case the peak can be constructed by drawing tangents along both slopes. The point of intersection is taken as a rough estimation of the point to which latency has to be measured. (see Fig. 3.10)

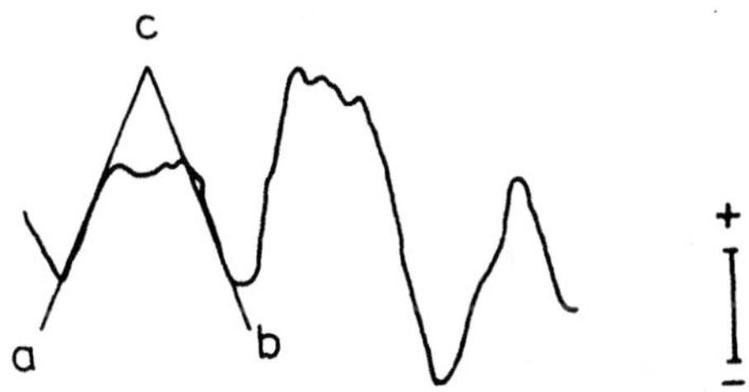

Fig. 3.10 Construction of the top of a flat peak.

In measuring the latency of waves IV and V there are two possibilities:
- either they form a complex in which the two peaks can easily be measured separately (IV-V complex)
- or they form a combined wave with one peak (IV-V wave).
It is advisable to determine normal values for both situations.

Amplitudes

These can be measured in two ways. First, the peak-to-peak values and second, the height of the peaks with respect to the average zero line. Due to the relatively high variability in individual subjects the peak-to-peak amplitude has a limited value as interpretation-parameter.
An usable parameter here is the ratio between the peak-to-peak values of wave V and I, called the V/I ratio. The peak-to-peak values of these waves are measured form peak to next negativity. (see Fig. 3.11B)

Polarity

All seven waves are vertex positive. There is however a difference between wave I and waves III to VII. Wave I and part of wave II are caused by the nerve action potential of the acoustic nerve. (see Fig. 3.2).

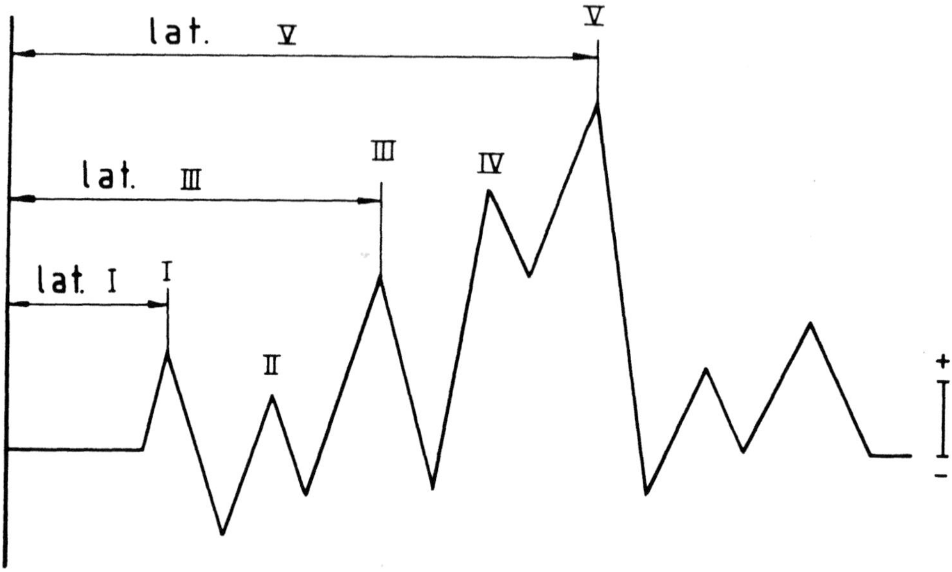

Fig. 3.11A Schematich drawing indicating how to measure the latencies.

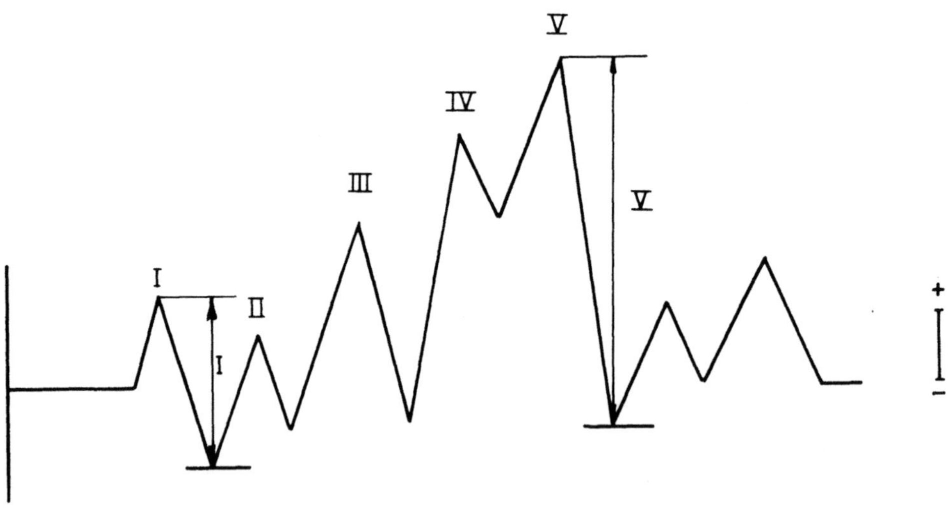

Fig. 3.11B Schematic drawing indicating how to measure the amplitudes.

This is a negative signal with the features of a near field potential. This potential influences Aipsi in a strongly negative way, while Cz becomes zero or slightly positive. So Aipsi is the most active electrode. Acontro is not influenced at all. If the vertex remains zero, wave I will not be visible in the derivation Cz-Acontro. If the vertex becomes slightly positive, an indication of wave I will be visible in the derivation Cz-Acontro.

Waves III to VII are far field potentials. They influence the vertex in a strongly positive way, while making Aipsi and Acontro slightly positive. This makes Cz the most active electrode here.

Wave II has an intermediate position between wave I and waves III to VII. As a result, all waves are recorded as vertex positive potentials, which by convention have their peaks upward.

Waves I, III and V as vectors

The most important waves of the BAEP are those numbered I, III and V. They may be conceived as separate vectors with a certain size and direction (vector is a more mathematical term, in physics it is called a dipole). The direction of the vector is important in determining the derivation that is to be used. The vectors of waves I and III are directed horizontally, that of wave V vertically. (see Fig. 3.12) The highest amplitudes of waves I and III can be measured by using the derivation Acontro-Aipsi, of wave V by using the derivation Cz-Aipsi. It is common practice to derive from Cz to Aipsi and from Cz to Acontro, as this produces the best results for the signal as a whole. But when wave I or III is indistinct, it is possible to deviate from this rule and to try the derivation Ac-Aipsi.

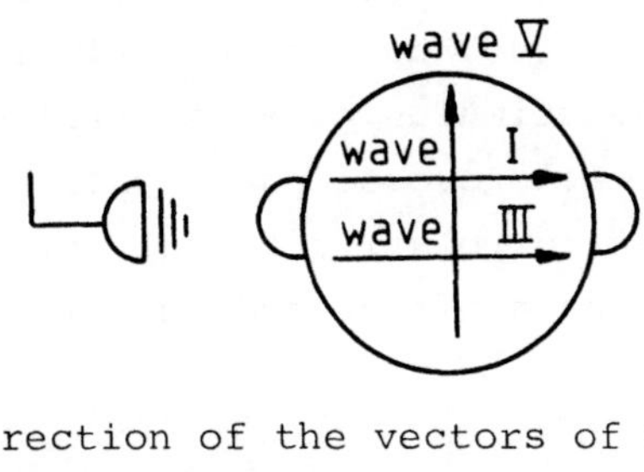

Fig. 3.12 Direction of the vectors of waves I, III and V.

Stimulus artifact and Cochlear Microphonics

Two more phenomena that one has to deal with and that may present difficulties for the interpretation are:

- Stimulus artifact. This occurrs practically always, though in varying degrees, and depends on stimulus intensity and duration. With the aid of a thorough electric and magnetic shielding of the headphone, the stimulus artifact can be limited somewhat, though it cannot always be prevented. A possibility here is the use of alternating stimulation (i.e. rarefaction and condensation). However, the normal values are then different.
- Cochlear Microphonics (CM). This wave can often be recorded just before wave I. It is one of the three potentials registered in the ECochG (electrocochleogram) and originates in the cochlea. It can be difficult to differentiate between the CM and wave I. This can be solved easily by reversing stimulus polarity. The CM will be reversed whereas wave I will remain in position.

4.1.6 Identification and differentiation of the waves.

Of course, the practice of the BAEP is more difficult than its theory. Waves will often be far less distinct than those in the nice picture in the book. As an aid in these practical problems a systematic discussion of ways to improve the forms of the waves follows.

To start with, it is often helpful to relax the patient or even sedate him. Furthermore, the number of samples per trial can be raised (4000 instead of 2000). Frequently a comparison of the derivation Cz-Aipsi with that of Cz-Acontro proves useful; so does a reversal of the stimulus polarity.

Lastly (mindful of the direction of the vectors), the derivation may be changed. The following applies to the separate waves:

Before wave I problems may arise, due to the stimulus artifact and the CM. Reversal of the stimulus polarity will help (stimulus artifact and CM will reverse phase whereas wave I remains the same). Alternating polarity of the stimulus is another possibility (stimulus artifact and CM will disappear through averaging).

Wave I:
- "horizontal" derivation (Acontro to Aipsi)
- insert an electrode deep into the meatus acusticus externus instead of into Aipsi (this is preferably done by an otoligist)
- increase stimulus intensity
- lower stimulus repetition rate
- pretragus instead of mastoid electrode position

Wave III:
- "horizontal" derivation (Acontro to Aipsi)
- increase stimulus intensity
- lower stimulus repetition rate
- reverse stimulus polarity
- adjust bandpass to 300-3000 Hz

Wave V:
- lower stimulus intensity (if necessary to 10 dB SL, in which all other waves will disappear)
- increase stimulus repetition rate
- reverse stimulus polarity

4.2 MLAEP

Methods

The equipment needed to record the MLAEP's, the electrodes and electrodeposition is essentially the same as that used to record the BAEP's. The analysis time is usually $\pm$ 100 ms, including 25% prestimulus time. The stimulation rate, therefore, cannot exceed 10 per sec. In all of the early studies of the MLAEP's, clicks were used as a stimulus. In the audiological application, tone pips, tone bursts and filtered clicks were applied in order to obtain frequency specific audiometric information. Comparing clicks and tonal stimuli, Zerlin (1973) found a more easily recognized NaPa response using click stimulation. The appropriate bandpass in MLAEP is 5 or 30 Hz to 250 Hz.

The effect of stimulus rate and intensity on latency and amplitude is moderate. Generally, latencies for the different components decrease with increased stimulation rate and tend to decrease with increased intensity. Increased signal intensity results in increased peak amplitudes (Geisler 1958, Mendel 1969, Gerin 1972, Thornton 1977, Wolf 1978).

The reproducibility of the MLAEP's within individuals and across time is good, in adults as well as in children. The responses are stable when the patient is awake and even during the different stages of sleep (Mendel 1969, 1974, 1975), and proved to be resistant to several hypnotic drugs (Mendel 1977).

4.3 AEP

When recording the cortical AEP it is often necessary to record more than one channel. For evaluation of general neurological properties of the response of the CNS to acoustic stimuli at least 3 channels are necessary, but more are preferable. Only when evoked response audiometry is performed is one vertex derivation sufficient.

4.3.1 Electrode positions

One electrode is placed at the vertex, and two are placed over the primary auditory areas left and right. When more electrodes are possible other positions over the temporal and central areas can be chosen. To achieve inter-patient comparibilit the normal electrode positions from the 10-20 system are preferable. Consequently, the temporal electrodes at T3 and T4 and Cz are the most important. Other positions are C3 and C4 and T5 and T6. A good response can also be obtained at P3 and P4. Electrodes can be referred to the linked mastoïds, linked earlobes, the common average reference electrode, or bipolar to Fz (from the temporal electrodes).

In our laboratory we use a row of electrodes from T3 to T4 referred to the common reference of Fz, with Cz referred to the earlobe. This leads to 10 channels to be analyzed. In Fig. 3.13 the configuration is given.

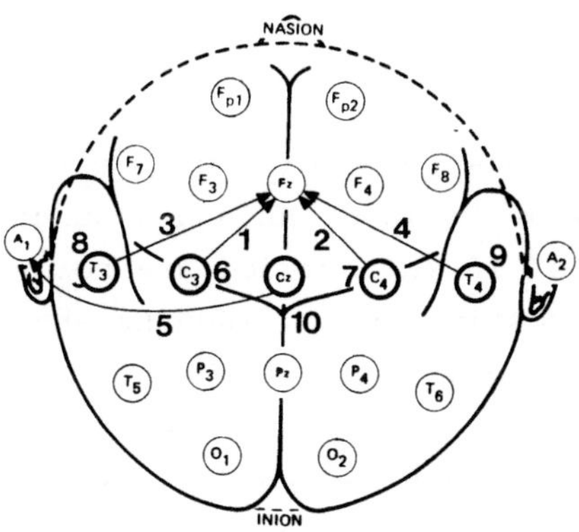

1 C_3 - F_o
2 C_4 - F_o
3 T_3 - F_o
4 T_4 - F_o
5 C_o - A_1
6 C_3 - Av
7 C_4 - Av
8 T_3 - Av
9 T_4 - Av
10 C_o - Av

Fig. 3.13 Example of electrode positions for recording the long latency AEP.

4.3.2 Instrumentation

For the recording of long latency evoked responses no special amplifiers are necessary. Normal EEG amplifiers with a low-pass filtering over 75 Hz are amply sufficient. A time-constant of 0.3 is normally used. For averaging, a sampling time interval of at least 5 ms is required. Higher sampling rates are better. In our laboratory 1 ms is the standard sampling time.

Special purpose averagers are now available which are convenient and sufficient for present clinical applications. Usually four channels are available and the choice of parameter settings is limited. Some can only be used for BAEP recordings, others offer a wider scope. General purpose systems using user-friendly program packages are gradually being offered. They provide many more possibilities, in number of channels, parameter settings, etc. They also offer a much wider choice of brain signal analysis, besides visual and somatosensory evoked response calculation.

When recording the long latency AEP, a prestimulus interval is advisory to assess background activity. The period analysed must not be shorter than 400 ms as in pathology latencies can be extremely prolonged. Stimuli have to be presented at random with intervals between 1.5 and 3 seconds to prevent habituation and accidental locking to background activity.

4.3.3 Stimulus parameters

For neurological investigations pure tones with a frequency of 1000 Hz, an intensity of 60 or 70 dB HL and a duration of 50 to 60 ms with ramptimes of 10 ms included are normally used. For ERA three frequencies (mostly 500 Hz, 1000 Hz and 4000 Hz) are used, with intensities around 30 dB HL. Because only detection of a wave related to earlier measurements is required, the number of stimuli is reduced,64 sweeps usually suffice. For neurological investigations usually both ears are stimulated with the subject relaxed as with normal clinical EEG-recordings. The normal procedures to prevent artifacts etc. have to be followed. For ERA several methods are available. The most frequently used technique is stimulation of one ear while the other is masked with white noise, and measurement of 4 to 5 EP's with decreasing tone intensity. Normograms are available (Sturzbecher et al. 1975) to assess the subjects hearing threshold. It is necessary for each laboratory to obtain values of a number of normal subjects in order to compare them with these normograms and to correct them for the spcific situation in the laboratory. This kind of callibration always remains a necessity.

Another method of performing ERA has been introduced by Campbell et al. (1977) using amplitude modulated pure tones or high frequency click-stimuli. With click-stimuli with a frequency of between 12-16 Hz a short test period is sufficient to evaluate the patients threshold within 5 dB. This method may seem advantageous compared to threshold determinations using the BAEP as it takes less time; it yields, however, essentially the same information.

It must be pointed out that ERA using long-latency AEP's is a tedious method which needs much expertise and should be reserved for those patients with whom no other method for

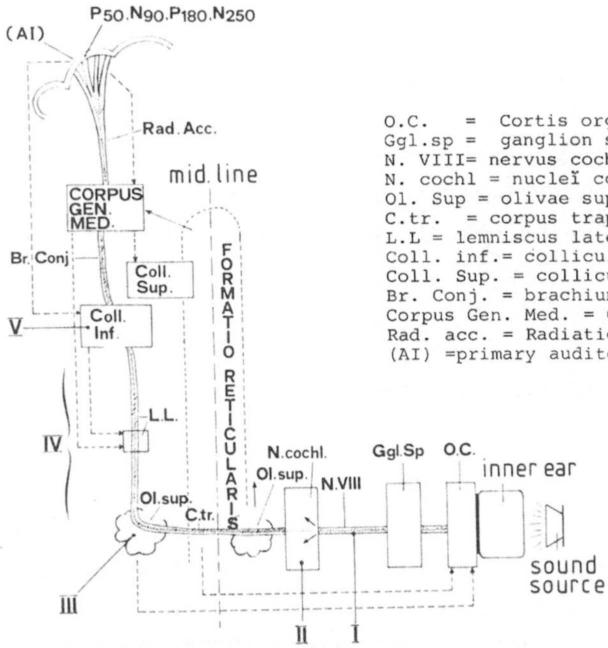

O.C. = Cortis organ
Ggl.sp = ganglion spirale
N. VIII= nervus cochlearis
N. cochl = nucleï cochleares
Ol. Sup = olivae superiores
C.tr. = corpus trapezoïdeum
L.L = lemniscus lateralis
Coll. inf.= colliculus inferior
Coll. Sup. = colliculus superior
Br. Conj. = brachium conjunctivum
Corpus Gen. Med. = Corpus Geniculatum Mediale
Rad. acc. = Radiatio Accoustica
(AI) =primary auditory cortex area A1.

Fig. 3.14 Schematic representation of the auditory system indicating the approximate place of origin of the main components. The roman figures represent the possible structures responsible for the generation of the corresponding BAEP components. It must be stressed that the actual visible components in the response are not caused by the neuronal activity within the indicated structures but rather due to the synchronized volley of depolarisations set-up in the efferent fibres. (i.e. wave II originates in the cochlear nucleï,but is due to the synchronous activity travelling along the corpus trapezoïdeus).

obtaining hearing thresholds is possible. Its main field of application will be in pediatrics for small children and in the evaluation of autistic children.

The averaged evoked reponse must always be written on paper, using some kind of plotting device. Use of a computer for calculation of positive and negative wave latencies and amplitude is recommended. The use of rulers may introduce errors, especially in latencies which make reliable left-right comparisons difficult.

5. LOCALIZATION OF THE VARIOUS COMPONENTS

5.1 BAEP

From section 2 and 3, the origin of the BAEP can be rather well established. Although some controversies still exists regarding this aspect, the following origins are posited by different authors, which also concur with animal data.
In Fig. 3.14 the origin of these components is shown. Wave I finds its origin in the auditory nerve fibres and wave II is the reflection of the synchronous activity in the cochlear nucleï. However, recent studies indicate that wave II can have its origin in the vestibular nerve. Regarding the third and fourth wave, the origin is less well defined. Presumably the former is generated by the information-processing in the olivary nucleus whereas the latter is the result of synaptic activity along the lateral lemniscal nucleï. The converging activity in the inferior colliculus is the source of wave V.

5.2 MLAEP

The origin of the MLAEP's

Considerable controversy exists with respect to the myogenic and cortical sources in the generation of the MLAEP. Geisler suggested a cortical origin, supported by others (Mast, 1965, Harker 1977, Pickton 1974). Bickford (1964) attributed the generation of the MLAEP to myogenic potentials of the scalpmusculature. Goff (1977) showed by simultaneous skull surface and intracranial depth recordings that the components of the complex had their origin from both myogenic and multiple neurogenic sources. The positive components with latencies of 25 and 50 ms and the negative component of 40 ms, the W-shaped P25-N40-P50 (PaNbPb) complex, according to Goff, are predominantly neurogenic. Study of the scalp distribution of the potential revealed an amplitude maximum at the vertex. The source of Pa lies approximately at the level of the Sylvian fissure (Cohen 1981). Hashimoto (1981) attributes the origin of No and Na to the midbrain, presumably generated as postsynaptic activity from the inferior colliculi. The early components of the MLAEP might therefore serve as an indicator for midbrain dysfunction.

6. DESCRIPTION OF THE NORMAL AUDITORY EVOKED POTENTIALS IN MAN

6.1 BAEP

6.1.1 Morphology

The BAEP occurs within 10 ms after the administration of a click. The curve consists of seven waves, designated I to VII (Jewett and Williston, 1971). Wave I has a latency of less than 2 ms; the following waves are found with intervals of about 1 ms. This means that the signal has a frequency of about 1000 Hz. Therefore a bandpass with limits of 150 and 3000 Hz is applied. This results in the exclusion of the EEG signal whereas the EMG signal can pass freely.

As a rule the waves have amplitudes of less than 0,5 mmV. As shown in fig. 3.15,wave I is usually observed in the ipsilateral derivation (i.e. at the side of the stimulated ear). Waves I and III are normally regular, well-developed waves, while the amplitude of wave III is significantly higher than those of waves I and II. Waves IV and V mostly form a complex, with their separate peaks clearly distinguishable. Sometimes, however, they form a combined wave IV-V of which only one peak is seen. This peak has a shorter latency than that of wave V of the IV-V complex (see Fig. 3.16).

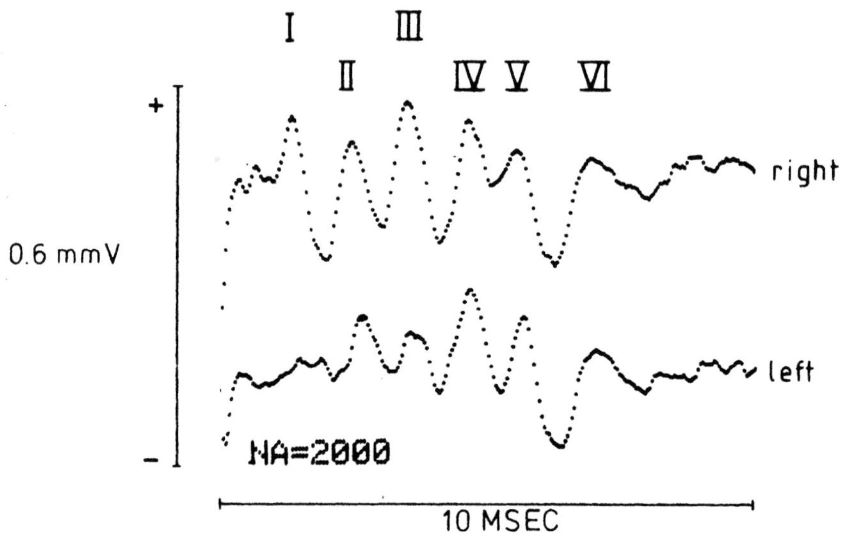

Fig. 3.15 Normal BAEP, stimulation right ear.

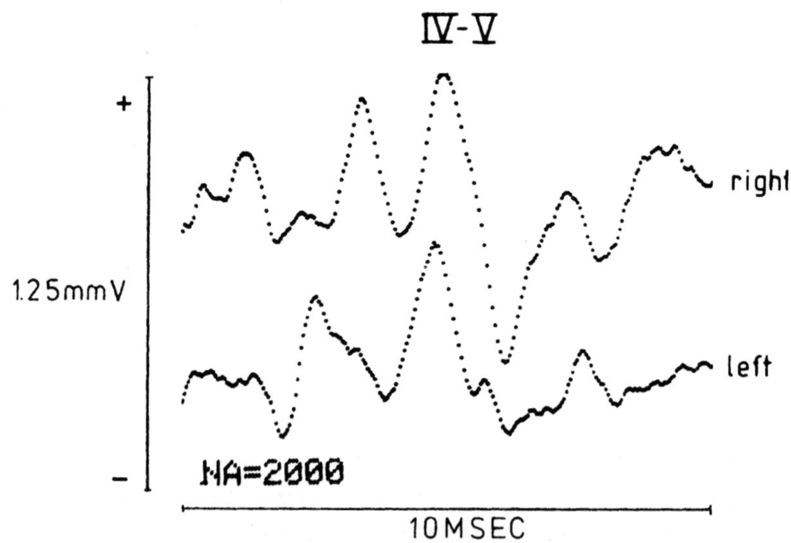

Fig. 3.16 Stimulation right ear. Combined wave IV-V.

The IV-V complex is very constant, as is wave III, and as a rule has the highest amplitude of the entire curve. Waves VI and VII have a greater variability with lower amplitudes. They are not very constant and they are sometimes absent (in normals!).

There are several differences between the ipsi and contralateral curves. In the latter, wave I is mostly absent and wave III is less developed than in the ipsilateral derivation. Moreover, the amplitude of wave III is considerably lower on the contralateral side, though the amplitude of wave II may be higher. Waves IV and V are mostly presented as a complex and hardly ever as a combined wave. When compared to the ipsilateral curve, their latencies are often longer.

6.1.2 Features

There are certain aspects in which the BAEP clearly distinguishes itself from other evoked potentials:

- The BAEP is registered within 10 ms after the administration of an acoustic stimulus (click).
- The amplitudes of the signal are usually lower than 0,5 microvolt.
- Its degree of reproducibility is very high, even within a span of months (low intraindividual variability). The interindividual variability in normals is also low.
- No habituation occurs.
- The signal is hardly or not affected by the degree of alertness of the patient (drowsiness, sleep, coma), or his medication.
- The signal has the highest frequency contents of all EP's.
- The BAEP makes it possible to examine not only the auditory function but also the function of the auditory pathways in pons and mesencephalon.
- Through stimulation with clicks the basal part of the cochlea is stimulated only.
- Examination by means of the BAEP is non-invasive, reliable and objective.

6.1.3 Parameter assessment

Peak latencies and interpeak latencies (PL's and IPL's)

The PL's of wave I, III and V are far and away the most important. Roughly, wave I represents the acoustic nerve, wave III the pons and wave V the mesencephalon. (Jewett, 1970, Buchwald and Huang 1975). I, III and V are usually guite distinct. The IPL's can easily be derived from their PL's, viz. the IPL's I-III, III-V and I-V. It is generally assumed that I-III gives information about the pons, III-V about the transition pons-mesencephalon and I-V about both. The IPL I-V is the most important. This interval is called the Central Conduction Time (CCT). (see Fig. 3.17).

As with all Evoked Potentials, each laboratory has to define its own normal values. It is all the more remarkable that the mean CCT has nearly everywhere been determined at 4,0 ms, for neonates at 5,0 ms.

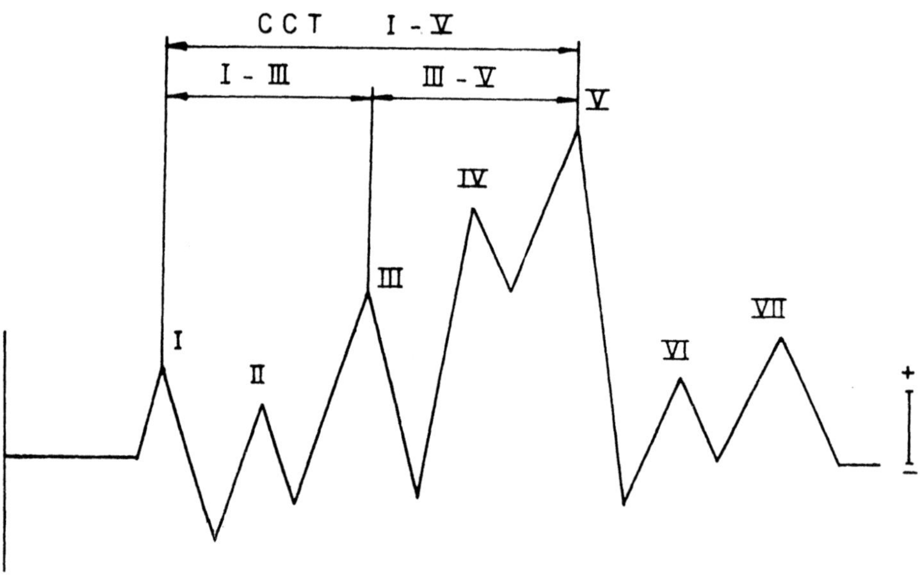

Fig. 3.17 Schematic drawing of IPL's.

Symmetry.

Several asymmetries may occur in a derivation from Cz to Aipsi and from Cz to Acontro.

Firstly, between Cz-Aipsi right and Cz-Aipsi left ; secondly, between Cz-Aipsi right and Cz-Acontro right or between Cz-Aipsi left and Cz-Acontro left.

A certain - though limited - asymmetry is allowed (cf. normal values).

Asymmetry between the PL of wave V right (ipsi-lateral) and left (contralateral) is most important, this must be less than 0.4 ms.

Amplitudes

The amplitudes are measured at the downslope of the waves. They do not serve as parameters in themselves, because their variability is too large. The amplitudo ratio V/I however is sometimes used as a parameter (i.e. the amplitude of wave V divided by that of wave I).

With normal stimulation the following values are used:

V/I ratio of > 1 is normal.

V/I ratio 0,5-1 is suspect.

V/I ratio < 0,5 is abnormal.

Opinions differ on the application of the V/I ratio as a normative parameter. In combination with a prolonged IPL the applications seems justifiable. But what if only the V/I ratio is abnormal? Should the curve be interpreted as abnormal? In our opinion it should not, especially because in normal subjects the V/I ratio if often lower than 1,0. Moreover the amplitude of wave V depends very much on the placement of the earlobe electrode (cf. relevant section).

Reproducibility

A BAEP has to be recorded at least twice in the same session in order to judge its reproducibility. Here questions arise as to the use of reproducibility as a normative parameter. On this problem too, opinions differ. We feel that it should not be used as such. Every BAEP, normal or abnormal, must be well-reproducible. If not, the cause may be an imperfect registration technique or a surplus of EMG artifact. In such cases either the technical deficiencies must be corrected or the patient sedated.

Relation stimulus - normative parameters

To achieve a proper understanding of the BAEP and its clinical applications in neurologic and audiometric research, a full appreciation of the relation between the features of the stimuli and those of the elicited waves is required. This is one of the most important points of the BAEP examination. Prominent features of the stimuli are polarity, stimulus rate and intensity (Stockard et al. 1978).

Polarity

We stimulate almost exclusively by means of rarefaction clicks.

Stimulus rate

In our clinic we use 11 Hz in neurologic and 31 Hz in audiologic examination.

Intensity

This is the most important stimulus parameter because it is often varied in the course of a recording session (as. opposed to frequency and polarity, which are kept constant as much as possible).

The most prominent features are:

- With an increase in stimulus intensity, all latencies decrease. IPL's, however, remain constant. The entire curve seems to shift to the left.
- With an increase in stimulus intensity, the amplitudes increase too. This applies only to a small extent to wave V.
- As follows from above, with a decrease in stimulus intensity wave V will disappear latest: it can still be registered at 5-10 dB SL.

In audiologic practice this information is used by setting up a diagram showing the relation between latency and intensity for one wave, usually wave V. This is called the latency-intensity curve. (see Fig. 3.18) The latency of wave V is registered with either increasing or decreasing intensity in steps of 10 or 5 dB. This produces a curve with a downslope to the right (the two quantities are inversely proportional). The decline of the curve amounts to about 40 mms per dB (limits 30-60 ms per dB). (Galambos and Hecox, 1978).

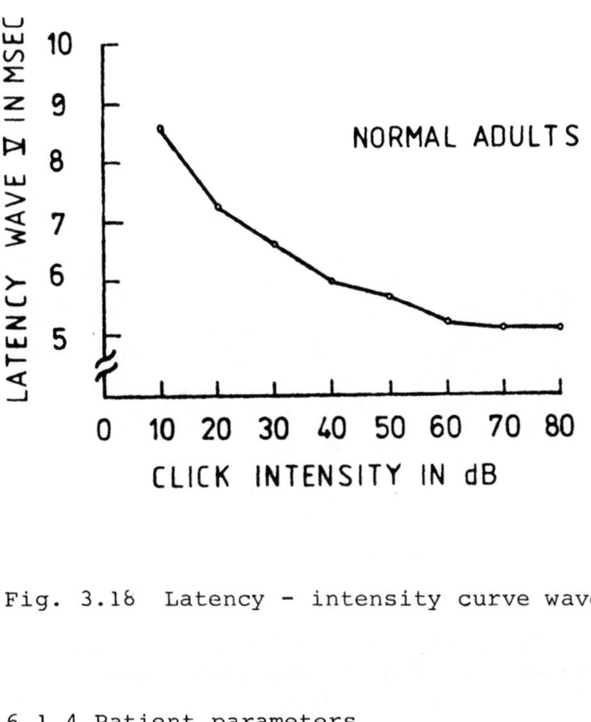

Fig. 3.16 Latency - intensity curve wave V.

6.1.4 Patient parameters

Most prominent are age, sex and body temperature.

Age

It is possible to register the BAEP in premature children with a conceptional age of 28 weeks (J.E. Stockard et al. 1981).

This BAEP is less differentiated with lower amplitudes and longer latencies. (see Fig. 3.19).

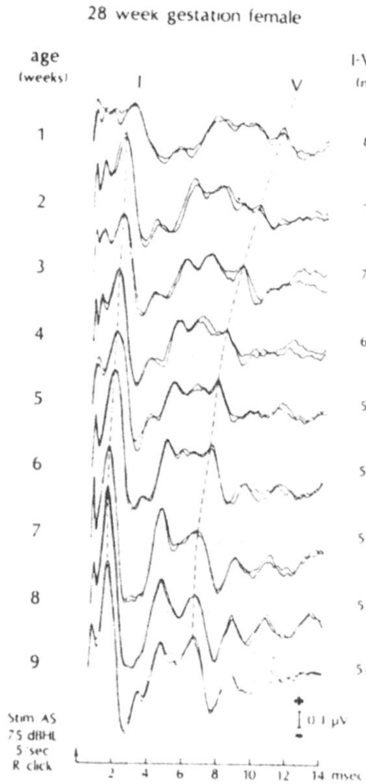

Fig. 3.19 Serial studies of normal 28-week-gestation infant. Over the nine-week period of study, wave I decreased by 1,36 ms and wave V by 4,78 ms. The I-V IPL decreased in a non-linear fashion, with greater weekly changes occurring in early prematurity (0,7 ms per week). Note that the latencies are far more stable over time as term is approached. Component identification was confirmed by simultaneous recording of vertex to contralateral ear. (reproduced from J.E.Stockard, publication: Tecnnical Considerations in the Recording and Interpretation of the BAEP for Neonatal Neurologic Diagnosis, 1981, with permission).

With growing age the BAEP develops too. At the age of 3, the BAEP is fully developed and comparable to that of an adult (Starr et al. 1977, Hecox et al. 1974).
The regularity of this phenomenon is such that the BAEP can be used as a parameter for cerebral maturation. It is in fact far more suitable than the EEG because of the higher variability in the latter (especially in young children). Normal values should be defined for the neonate at the ages of 3, 6, 9, 12 and 24 months.
Several researchers have described the prolongation of the BAEP latencies in subjects over 40 years. We could not confirm that by our own experiences.

Sex

It is often mentioned that women have shorter latencies than men. However, we did not find these differences.

Temperature

Latencies increase with the lowering of oesophagal temperature. Under such circumstances however, the temperature must be extremely low, as might happen in a prolonged operation with hypothermia. In practice this seldom occurs.

6.2 MLAEP

Recently the middle latency evoked potentials (MLAEP), occurring 8-100 ms after stimulation, have also received attention.
The human MLAEP was first described by Geisler (1958). Because of the controversy about the myogenic versus neurogenic origin of the MLAEP the clinical application of MLAEP has been very limited.
The latency range of the MLAEP is 8 to 100 ms. It coincides in part with the range of myogenic potentials. Those myogenic potentials have considerably higher amplitudes and can be avoided by artifact rejection of wave configurations with amplitudes above 9 mmV. The potential consists of polyphasic positive waves (Po,Pa,Pb,Pc) and negative troughs (No,Na,Nb,Nc), according

to Pickton. The No or the PoNoPa components are identical with the SN-10 of the BAEP. The SN-10 occurs at a peak latency of 9-10 ms. (Davis, 1979) These early components are followed by the most prominent trough Na (24-28 ms), the positive peak Pa (30-40 ms) and negative trough Nb (40-47 ms). (see Fig. 3.20).

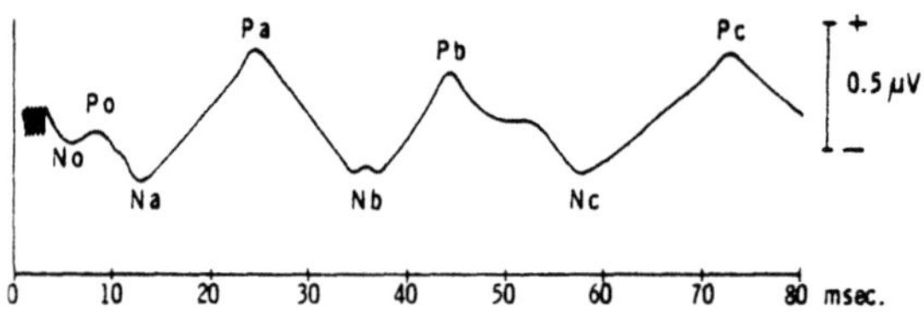

Fig. 3.20 Schematic representation of a MLAEP in an adult.

Recordings in adults usually do not show significant lateralisatio in latency or amplitude after monaural stimulation. In infants, however, the NaPaNb components are recognized predominantly on the stimulated side (Fig. 3.21).

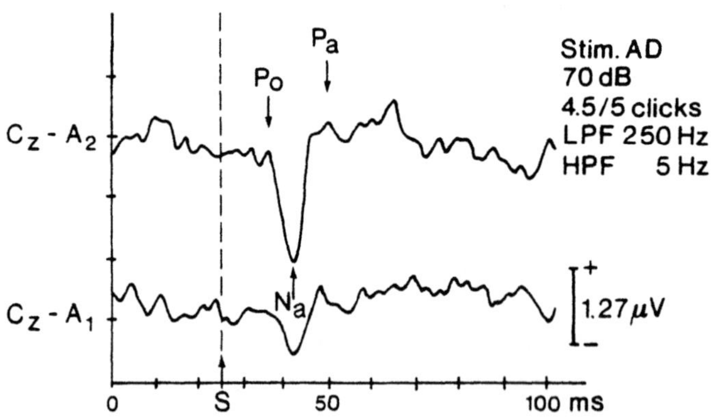

Fig. 3.21 MLAEP in a 2 weeks old normal mature female infant. Po = 10.6 ms, Na = 16.6 ms and Pa = 20.7 ms Stimulation right ear, 512 clicks 70 dB, BP 5-250 Hz.

6.3 AEP

6.3.1 Response properties of the AEP

The BAEP is described separately. The amplitudes of the later components of the AEP are much larger than the earlier ones. For the middle components amplitudes are only slightly larger than the BAEP amplitude levels, but for the cortical AEP amplitude values are several microvolts to 10 - 20 microvolts for components between 100 and 300 ms after the stimulus. The middle components have latencies between 10 and 50 ms, the latest component being identical to the first one of the late cortical response. The middle components can be recorded with approximately 100 to 200 stimuli, mostly clicks, using a sampling rate of 1000 Hz. It is important to use analog filters to suppress the other components of the AEP. The best filter setting is a bandpass filter with settings between 25 and 175 Hz (McFarland et al. 1975). With these precautions the middle latency AEP can be detected. It is said not to habituate (Vivion et al. 1977) contrary to the late AEP. The origin of its components is not clear.

The cortical AEP to long interval click or tone stimuli consists of 5 to 6 dominant waves designated P1, P2, P3 and N3 of which, for clinical use, the N1 (with a latency of around 100 ms), the P2 (with a latency between 180 and 200 ms) and N3 (around 300 ms) are the most important deflections seen in the vertex recorded AEP. When recording from temporal electrodes the polarity is reversed. (Davis et al. 1966) (see Fig. 3.22)

The P1 has a latency between 50 and 60 ms and other small peaks between 20 and 50 ms may precede the P1. These waves are not seen if a FM-stimulus is used. With the latter three clear waves can be detected which have the same latency range as is demonstrated for click stimuli. With ramp durations of 20 ms and a modulation of 200 Hz at 1000 Hz base frequency, stable N1 and P2 latencies and amplitudes are found. Latencies increase and amplitudes decrease with intensities lower than 40 dB. Between 40 and 60 dB HL the response is stable. Identification is done based on latency and amplitude values obtained from series of normal subjects which are published in the literature

(e.g. Davis et al. 1964, Bostock and Jarvis 1970, Pickton et al. 1974). The latency and amplitude values from different authors are given in table 3.1; only the values from different latency components are given in this table. The late AEP habituates and is also influenced by the attentive state of the subject.

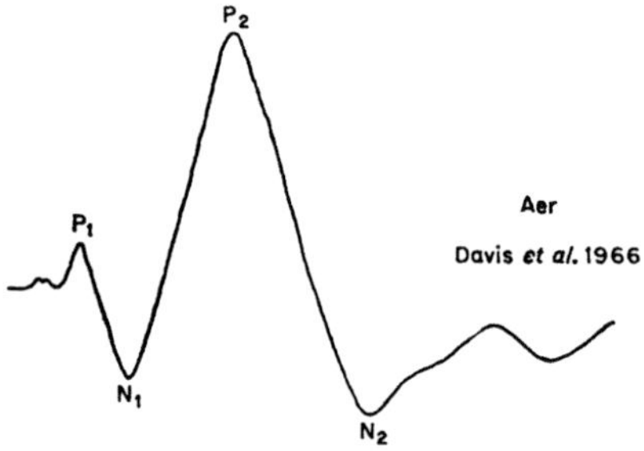

Fig. 3.22 Nomenclature for the main components of the AEP most frequently used. Depending on the polarity of the input channels the components are either depicted with the positivity upwards (as given in this original published by Davis, et al. in 1966) or with the normal EEG conventions.

6.3.2 Parameter assessment

The latencies are the most important parameters of the cortical AEP, especially with regard to their symmetries. Asymmetric latencies over 5 to 10 ms - depending on the wave - are always suspect for pathology. As has been mentioned before, the most important waves are the N1 and P2 for the response both to amplitude-and to frequency-modulated stimulation. Response latencies more than two standard deviations longer than the normative values are considered abnormal.

Latencies are always measured from the peak of the wave and referred to the stimulus onset.

The amplitudes are much less reliable as parameters for the assessment of abnormality in neurological applications. For ERA the absence of a complex is the criterium for threshold determination, although in another approach extrapolations, using normograms, can be used.

Asymmetries in amplitude have little clinical relevance. Only asymmetries over 60% may bear some significance. Usually latency differences are also present which are more reliable.

It is necessary always to use the same stimulus parameters as response has a rather stable relation hereto. Only those stimuli for which laboratory standards are established, must be used clinically.

6.3.3 Reproducibility

The reproducibility of the cortical AEP is very much dependent on the attentive state of the subject. Intra individual reproducibility is high,especially when two recordings are done successively and the subject remains alert. The same applies for longer intervals; however, differences in electrode position and attentive state may degrade the results. When recording AEP it is recommended to record always a response without stimulus, a so-called 'deaf' AEP, for comparison. Especially when components are difficult to detect this may be of importance.

6.3.4 Patient parameters

As has already been mentioned, changes in attentive state are the major sources of variability for the cortical AEP. Another source is age. Latency and amplitude values are influenced by age. There are only a few studies available assessing the development of the AEP.

On smaller children an increase in amplitude and decrease in latency is found (Ornitz et al. 1972). For adults age has practically no influence. The cortical AEP can be recorded in prematures and may be used in the assessment of conceptual age.

7. NORMATIVE VALUES

7.1 BAEP

7.1.1 Introduction

The latencies of the wave I up to and including wave VII, and the interpeak latencies calculated from these depend on several factors.
These factors can be divided into:

a. Technical factors, like equipment and parameters for stimulation and registration.
b. Physiological factors,like age, sex and body temperature.

In practice many laboratories use different stimulus parameters and different equipment as well in registering the BAEP. This implies that each laboratory must determine its own normative values. However, this if often difficult to realize. In these cases a practical solution can be found in taking the normative values of another laboratory as a starting-point and in choosing the same stimulus parameters. With these identical stimulus parameters, tests are performed at random in several groups of subjects. When the values so found are in accordance with the normative values of the model laboratory, these values may be used for clinical application.

7.1.2 Our own normative values

In determining the normative values of the BAEP one may use relatively small groups of subjects because there is hardly any variation. We used four groups:

A. women older than three and younger than 40.
B. men older than three and younger than 40.
C. women older than 40.
D. men older than 40.

The total number of subjects was 32 (4 x 8) persons, non of whom had ever suffered from audiologic an/or neurologic disorders. Furthermore, the disparity between the hearing threshold right and left was not allowed to exceed 5 dB.
We proceeded to determine:

a. the latencies of all peaks ipsi-and contralateral of the

stimulated side.

b. the interpeak latencies ipsi- and contralateral of the stimulated side.

According to these data we concluded that:

- no significant age-dependence exists for the peak latencies and the interpeak latencies in persons older than three years of age.
- no significant difference exists for the peak latencies and the interpeak latencies between men and women. (see Table 3.1).

Wave	Ipsilateral	Contralateral
I	1,67 ± 0,13	-
II	2,71 ± 0,19	2,84 ± 0,17
III	3,76 ± 0,21	3,65 ± 0,22
IV	4,98 ± 0,22	4,90 ± 0,20
V	5,64 ± 0,17	5,75 ± 0,20
VI	7,24 ± 0,37	7,35 ± 0,39
VII	8,82 ± 0,41	8,71 ± 0,47
I - III	2,08 ± 0,13	-
III - V	1,89 ± 0,15	2,08 ± 0,15
I - V	3,97 ± 0,15	-

Table 3.1 Normative values of 32 subjects (the total of the groups A.B.C. and D).

With respect to the symmetry there are two possibilities:

1. Firstly, an asymmetry between the peak latencies and interpeak latencies ipsi- and contralateral of the stimulated side, also between Cz-Aipsi and Cz-Acontro.
2. Secondly, an asymmetry between the latencies of the peaks and interpeak latencies ipsilateral in right and left stimulation, so between Cz-Aipsi right and Cz-Aipsi left. (see Fig. 3.23).

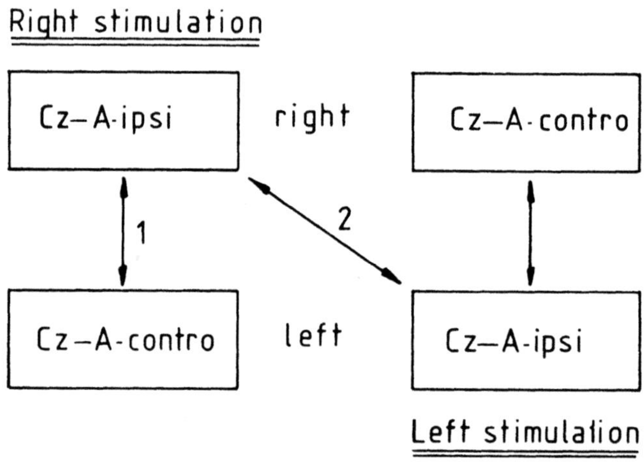

Fig. 3.23 1. asymmetry between Cz-Aipsi and Cz-Acontro.
2. asymmetry between Cz-Aipsi right and Cz-Aipsi left.

In the latter situation we investigated the difference between the Central Conduction Time (=CCT =IPL I-V) in right and left stimulation (see 2 in Fig. 3.23).
This difference amounts to an average of 0,102 ms (SD = 0,1057). By using a borderline of + or - 3 x the standard deviation, this means that an asymmetry of more than 0,4 ms should be considered pathological.

7.1.3 Comparison of the normative values

When the peak latencies of the waves I/VII in persons above three years of age are compared, it appears that rather great differences exist between different laboratories. These are greatest in the later waves and often amount to 2 or 4 times the standard deviation.
They are for the most part caused by distinct methods of stimulation as well as differences in equipment (see Table 3.2).

7.2 MLAEP

ML-AEP Components			No	Po	Na	Pa	Nb	Pb	
Adults									
Mendel	1969	ms	7.4-7.9	16.7-16.8	26.8-27.1	40.2-41.3	52.9-55.0		50 dB click; BP:10-100/150 Hz
Mendel	1977	ms			19.1	29.2	39.4		60 dB 1000 Hz toneburst BP:10/500; 25/175
		muV			0.67	0.83	1.05		
Goff	1969	ms		(P1a)10	(N1b)13	(P2a)27	(N2b)35		85 dB click; n=128-512
Picton	1973	ms	8.9	12	16	25	36		60 dB click
Vivion	1977	ms		9.08-10.43	16.14-17.32	24.78-25.52	33.80-34.90	44.64-45.57	30/50 dB click n=128-512 100 mus 1,2,4,8,16/s
Musiek	1981	ms				27.5-33.0		50,0-62,0	BP:30-100 Hz 10/s, n=2048
Cohen	1981	ms			19.5	28.1(BP70/250Hz) 32 (BP10/250Hz)	41.7		50dB click; 11/s
Neonates and infants									
McRandle	1974	ms		11.0±6	17.2±5.2	28.2±5.4	35.2±2.1	51.1±5.0	55 dB click; 4.5-9/s BP:20-125Hz
		muV		PoNa:0.6±0.2	NaPa:1.2±0.6	PaNb:1.7±0.9	NbPb:1.4±0.7		
Mendel	1977	ms			16.5	24.2	37.6		60 dB toneburst BP:10/500; 25/175 Hz
		muV			0.45	0.54	0.38		

Table 3.2: Summary of different reference values for the ML-AEP components. BP=Band Pass; n=number of stimuli.

Table 3.2

When the interpeak latencies I-III, III-V and I-V are compared, the disparities turn out to be smaller: as a rule, smaller than one standard deviation. Especially the disparities in the central conduction time (I-V interval) are remarkably small. Frequently an average value of exactly 4,0 ms is established. (see Table 3.3).

	I - III		III - V		I - V	
	$\bar{X}$	SD	$\bar{X}$	SD	$\bar{X}$	SD
Stockard (1977) 60 dB SL 10 Hz	2,1	0,2	1,9	0,2	4,0	0,2
Wielaard (1981) 70 dB SL 15 Hz	2,19	0,12	1,85	0,16	4,04	0,20
Zonneveldt (1981) 70 dB HL 11 Hz	2,08	0,13	1,89	0,15	3,98	0,15
Jones (1982) 80 dB HL 10 Hz	2,12	0,20	1,82	0,14	3,95	0,18
Eggermont et al.	-	-	-	-	4,0	0,1
Manhand et al.(1982) 70 dB HL 10 Hz	2,17	0,17	1,94	0,15	4,08	0,21

Table 3.3 Normal values IPL.

The normative values of neonates are established by J.E.Stockard and Westmoreland (1981).
Below we give their figures as published in their article.
Their stimulus rate was 20 Hz (rarefaction clicks).
(see Table 3.4) .

J.E. Stockard et al., 1981.
Normative Data

		Peak latency			interpeak latency		
		I	III	V	I-III	I-V	III-V
	70 dB HL	1,81	4,62	6,72	2,78	4,90	2,15
	(n=62)	0,22	0,29	0,32	0,21	0,27	0,23
	60 dB HL	2,15	4,86	6,93	2,72	4,79	2,07
	(n=30)	0,21	0,33	0,32	0,20	0,20	0,22
38-42 weeks	50 dB HL	2,47	5,20	7,23	2,73	4,76	2,03
conceptional age	(n=30)	0,22	0,32	0,36	0,21	0,28	0,26
	40 dB HL	2,83	5,59	7,62	2,75	4,79	2,04
	(n=30)	0,26	0,33	0,42	0,23	0,32	0,23
	30 dB HL	3,45	6,07	8,10	2,62	4,65	2,02
	(n=30)	0,38	0,41	0,46	0,30	0,30	0,25
36-37 weeks	70 dB HL	1,80	4,80	7,07	2,99	5,27	2,27
conceptional age	(n=20)	0,12	0,16	0,27	0,19	0,31	0,29

Table 3.4

As an example of normative values in infants we give below the figures as established by Hecox (1980). It should be noted that a stimulus frequency of 30 Hz was used (alternating clicks). (see Table 3.5)

HECOX, K.E. (1980) Pediatric norms for monaural click stimulation (90 dB SPL at 30/s)

	neonates	6 weeks	3 months	6 months	9 months	1 year	adult.
Wave I	2,5	2,2	2,1	1,9	1,9	1,8	1,8
Wave III	5,1	4,8	4,6	4,4	4,3	4,0	3,9
Wave V	7,6	7,1	6,9	6,5	6,3	6,0	5,9
I - V	5,1	4,9	4,8	4,6	4,4	4,2	4,1
I - III	2,6	2,6	2,5	2,5	2,4	2,2	2,[illegible]
III - V	2,5	2,3	2,3	2,1	2,0	2,0	2[illegible]

Table 3.5

7.3 AEP

In this section the normative values will be given for the cortical AEP for tone and FM-stimuli and the ERA normative values as produced in the literature.

As it is not feasible to give all descriptions of normative values, only the most relevant are given, either depending on the series size or the accuracy of the methods used.

It is important to point out that one must be very careful in the interpretation of the results obtained with EP recordings. In a study on interjudge and intra-judge reliability, especially on the presence or absence of a response, Rose et al. (1971) found rather large differences. This is especially the case with threshold determinations and when no strict criteria are available. In another study (Blom et al. 1980) a rather good interjudge agreement was found for the assessment of components in the VEP. An essential tool is the use of a 'blind' EP to assess the contamination by background EEG and the availability of criteria. Computer assisted peak identification is another tool to diminish subjectivity (Blom et al. 1975).

In Table 3.7 normative values are given from various authors for the tone and the FM-stimulus. There is a rather good agreement in the latency value for the different components. Reviewing the literature it can be calculated that most standard deviations are around 20 ms in normal healthy subjects. In pathology most latencies are prolonged with 40 ms or more.

In fig. 3.24A latency curves are designed according to the best fitting curve for ERA as given by Stürzebecher et al. (1975) for 1000 Hz. In this article the best fitting curves for 4000 and 500 Hz are given as well. As for ERA, the amplitude is an important parameter; the N1-P2 relative amplitude curve is given in figure 3.24B for 1000 Hz.

The mathematical expressions are given in the captions of the figures. These values coïncide with other data in the literature. These authors show no relation with age for adults.

It is advisable to assess the normative values for one's own laboratory when introducing AEP into clinical practice, and to compare these with data from the literature. Sample sizes between 10 to 20 subjects are sufficient. A thorough audiologic examination is necessary to ensure normal hearing. Strict adherence to standard procedures is essential to obtain reliable results.

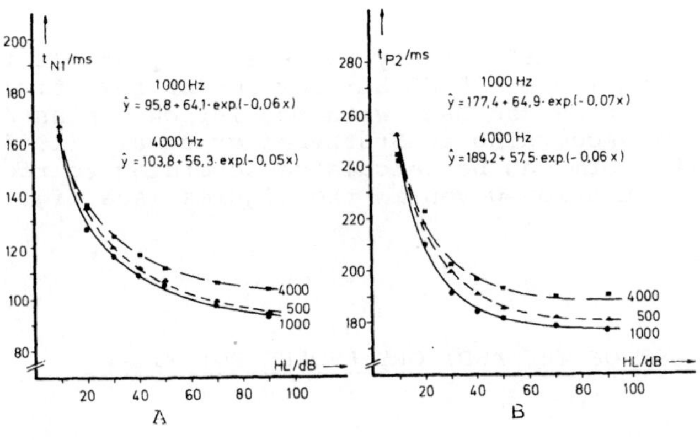

Fig. 3.24 Normative values for the latency of N1 (A) and P2 (B) calculated for normal adults for three different frequencies. The mathematical expressions are given in the figures. (reprinted with permission from Stuřzebecher et al. 1975)

	P_1	N_1	P_2	N_2
Tone AER				
Davis and Yoshie (1963)	50-60	95-105	170-200	300
Weitzman and Kremen (1965)	50	100	175	325
Gross et al. (1965)	100	160	220	-
Ornitz et al. (1967)	-	100	220	310
Bostock and Jarvis (1970)	60	110	180	260
Deeche et al. (1977)	36	91	177	318
FM-AER				
Lenhardt (1971)	-	-	-	
Arlinger (1976)	-	130	228	

Table 3.6 For tone-AER latency is relatively stable for stimuli over 25 dB HL; for FM-AER ramp times of <50 ms are used with modulation depths over 10%. According to Stürzebecker et al. (1975), the norm can be calculated according to the calculation given in the figures (see fig. 3.24).

8. CLINICAL USE OF THE AUDITORY EVOKED POTENTIALS

8.1 BAEP and ERA

8.1.1 Distinction between audiologic and neurologic disorders.

Three factors are essential for registering a normal BAEP:

a. an unimpaired hearing organ.

b. an unimpaired acoustic nerve.

c. an unimpaired lower brainstem.

Whenever the function of one of these structures is effected, an aberrant BAEP will occur.

The latency of wave I (a) represents the time needed to pass the hearing organ and the acoustic nerve. The time interval between wave I-IV, the CCT (central conduction time) (b), marks the time the signal needs to travel through the pons and mesencephalon. (see Fig. 3.25).

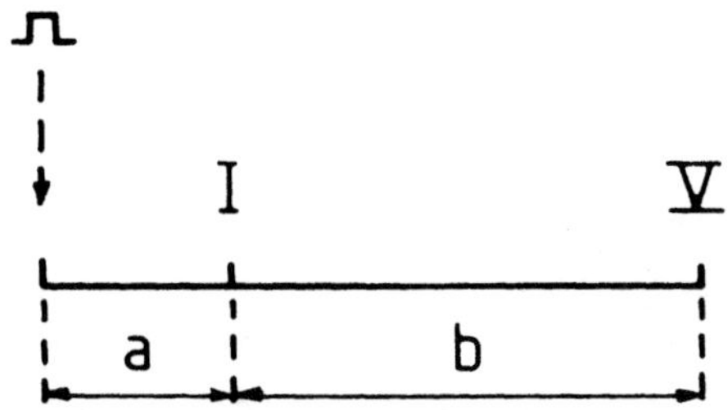

Fig. 3.25 a. stands for the time the signal needs to pass the hearing organ and the acoustic nerve.
b. stands for the time needed to pass the lower brainstem (notably those parts of the brainstem called pons and mesencephalon) (schematic).

In the case of hearing disorders, the latency of wave I will increase. This is due to an encumbered passage of the signal through the structures concerned. This means that in hearing disorders the latency of wave I should be determined, as it is done with the ECochG. Wave I of the BAEP can be compared to the action potential of the N1 of the ECochG.
The latency of the AP (N1) may vary between 1 and 4 milliseconds in normal ears and serves as a measure for the function of the auditory organ. A disadvantage inherent in wave I is that it vanishes fast when hearing acuity diminishes. Wave V, however, is already distinguishable with a very low stimulus intensity. In the case of hearing disorders the CCT remains about constant. When the latency of wave I increases the latency of wave V will increase about as much. We therefore see that in practice not the latency of wave I but that of wave V is used as a parameter in determining the auditory function.

In neurologic disorders the latency of wave I is not essential since this may be either normal or prolonged. Due to this aptness to vary it cannot be used as an interpretation parameter. The interpeak latencies, however, are important. When these are prolonged this points to an encumbrance in the passage of the signal in pons and/or mesencephalon. Relatively little attention is given to the absolute value of the latencies.

8.1.2 BAEP and audiometry.

ERA - Evoked Response Audiometry.
The purpose of this examination is to determine objectively the hearing threshold and hearing function, as opposed to ordinary audiometry in which cooperation of the patient is necessary.

The cortical auditive evoked potential (AEP) is the oldest technique used for providing information about the auditory function by means of ERA. One of its disadvantages is the large measure of intra- and interindividual variability. Another important drawback is the fact that in AEP registration the cooperation of the patient is indispensable, while at the same time his level of consciousness and possibly his medication are influential factors.

Afterwards the ECochG was developed which, contrary to the AEP, turned out to be a stable and reliable method. This, however, has the drawback of being an invasive method: the tympanum must be perforated by means of a needle electrode. The BAEP, on the other hand, has all the advantages of the ECochG while avoiding its disadvantage, since the BAEP is a non-invasive method.

The hearing threshold can be determined with the help of a latency/intensity curve of wave V. The lowest intensity which still gives a wave V (minus 5 to 10 dB) is the hearing threshold. This increases with the advance of age, but it may also be heightened in comparison with other subjects of a certain age group.

There are two kinds of deafness, conduction and cochlear deafness. This is connected with the fact that the ear consists of three parts:
- the outer ear.
- the middle ear.
- the inner ear.

The outer and middle ear conduct the sound waves to the inner ear. In the inner ear the perception takes place. It is a transducer which transmits the sound waves into nerve action potentials.

When there is a disturbance of the outer or middle ear, we speak of conduction deafness. A lesion of the inner ear causes a cochlear deafness, also called perception deafness.

The examination of the hearing acuity is performed with stimuli of 30 dB HL and a rate of 31 Hz. When under these circumstances wave V is present, the auditory function is such that normal social contact is possible. In differentiating between conduction and perception deafness the latency/intensity curve is also used. Both forms of deafness show a different curve, each with its own features. In the case of conduction deafness the curve shows a normal slant but has been moved in its totality to the right. In the case of cochlear deafness the curve has a more pronounced slope as a result of "recruiting" in the cochlea. (Galambos et al, 1978). (see Fig. 3.26). (reproduced with permission).

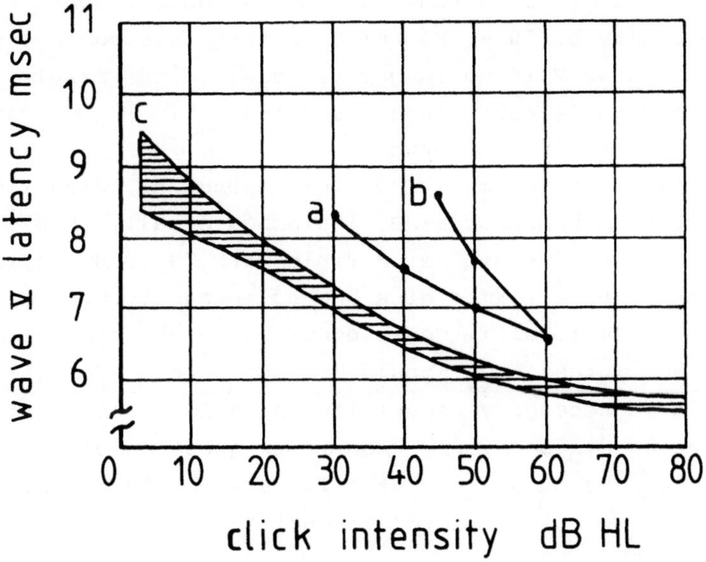

Fig.3.26 Differences of the latency/intensity curves of wave V in the case of: a. conductive hearing loss. b. sensorineural hearing loss (perception deafness) c. normal range.

In deafness due to a disturbance in the cortex,a normal BAEP and a disturbed or absent AEP are registered.

8.1.3 Standard procedure determination of the hearing acuity

The same electrodes are used as described in section 3, deriving form Cz to Aipsi. The stimuli consist of rarefaction clicks with a duration of 100 microseconds and a intensity of 30 dB above hearing level (HL). The stimulus frequency amounts to 31 Hz. Each trial is performed twice. The latency of wave V is taken as parameter.

When there is a distinct wave V, the hearing acuity is normal. When wave V is absent, it is necessary to increase the stimulus intensity with steps of 10 dB, till wave V appears. Then it is possible to define the sensation level more accurately by using steps of 5 dB. It is not necessary to measure the peak latency of wave V. Most important is the presence or absence of this wave.

In more detailed examination the latency/intensity curve of wave V is constructed. This means that we begin with a stimulus intensity of 70 dB HL which is than lowered in steps of 10 dB until wave V is no longer visible. In the resulting curve, the following is taken into account:

- the form of the curve.
- its slope (normally 40 microseconds/dB, with limits of 30-60 microseconds/dB) (Galambos et al., 1978).

When wave V is not perceptible with 70 dB HL, stimulation can safely be performed at a higher intensity with a maximum of 120 dB (clicks!!). Cases occur in which stimulation with even higher intensities should be considered, e.g. till 140 dB HL. This is necessary especially in children suffering from severe hearing dysfunctions, when it is questionable whether a hearing-aid can be of any use at all. Care should be taken not to stimulate at pain level intensities.

8.1.4 Advantages and disadvantages

Advantages:

- the patient's cooperation is not indispensable in this examination (which is often performed with children who may be sedated or even examined under anaesthesia).
- it is a well-standardized examination that gives objective information (ordinary audiology has subjective aspects).
- it is possible to differentiate between conduction- and cochlear deafness.

Disadvantages:

- the click consists mainly of high frequencies. This means that in audiometric examination only the basal part of the cochlea (where the high tones are received) is tested. Consequently, no information is given on the apical part. (Pratt and Bleich, 1982).
- it is no test of hearing itself. The hearing is a function of the higher centres in the cortex. The BAEP tests only the peripheral part of the hearing system. (It is obvious that this must be intact so as to provide these higher centres with information).

8.1.5 Indications

- objective estimation of the hearing level.
- children with maturation/development disturbances in which it is necessary to differentiate between auditory dysfunctions on the one hand and infantile autism, afasy and oligofrenia on the other.
- children with perinatal cerebral damage, chromosomal and genetic dysfunctions etc. (auditory dysfunctions occur frequently in this group).

Literature:

We have already mentioned the publication of Galambos (1978). See further: Brackmann (1977) and Skinner and Glattke (1977). The problems of ERA are also to be found in a book by Gibson and Ruben (1978).

8.2 BAEP and Neurology

8.2.1 The pathological BAEP

As mentioned before, alterations in the BAEP may differ widely, depending on their audiologic and neurologic origins:

a. audiologic disorders

The latencies of wave I and all successive waves show about the same increase so that the IPL's remain practically the same. The entire curve shifts to the right. In serious hearing disorders wave I may be completely absent.

b. neurologic disorders

The latency of wave I may vary. Its presence does, however, point to the existence of a certain degree of functioning of the hearing organ. Due to an encumbrance in the passage of the signal in the auditory pathway, the IPL's lengthen to a greater or lesser degree. This gives the impression of the whole signal being drawn apart. In more serious lesions one or more waves will disappear completely, either early or late ones. Contrary to what happens in auditory dysfunctions, wave I is usually the one which is visible the longest. (see Fig. 3.27).

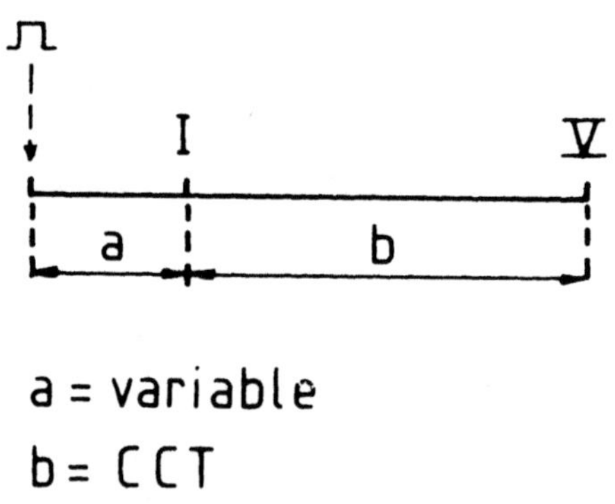

Fig. 3.27 Schematic drawing of the conduction time in neurologic disorders.

The most important feature is the lengthening of the CCT. The right-left disparities should also be considered (these should not be more than 0,4 ms, even in the case of normal time values). It is sometimes possible to make further differentiations:

I - III prolonged III - V normal	lesion in pons.
I - III normal III - V prolonged	lesion in mesencephalon.
I - III prolonged III - V prolonged	lesion in pons and mesencephalon.

Complete absence of the BAEP is the greatest abnormality. In neurology this may be seen in patients with brain death. The BAEP may also be absent in pre-existing double-sided deafness or in a double-sided fracture of the os petrosum. This is why many researchers demand that an absent BAEP be used as an argument in establishing cerebral death <u>only</u> when wave I is still clearly visible.

8.2.2 Indications

a. <u>processes in the posterior fossa</u>.

Either <u>outside</u> the brainstem (cerebello-pontine-angle tumor, tumor cerebelli) or <u>in</u> the brainstem (tumor, vascular disorder, encephalitis, syphilis, trauma, anoxia). The BAEP is the foremost diagnosticum in the case of a cerebello-pontine-angle tumor!

b. <u>demyelinisating diseases</u>.

Here multiple sclerosis is, of course, the most important. It is responsible for 30% of the total number of BAEPs that show disturbances.

c. <u>comata</u>.

We ust differentiate between a functional disorder (in which the BAEP is usually normal) and an anatomical lesion (in which the BAEP may be disturbed).

d. intoxication.

Most important are the acute and chronic alcohol intoxications.

e. traumata.

Concussion of the brainstem

f. cerebral death.

The waves disappear successively, starting with wave VII, then VI and so on. Wave I is the one longest present.

g. in neuropediatry.

The BAEP serves to judge the cerebral maturity of very young children (below the age of 3) with developmental disturbances.

h. in neurosurgery.

The BAEP serves as a monitor in operations of the posterior fossa, e.g. of a cerebello-pontine-angle tumor or neuralgia of the nervus V.

Below we will discuss the indications a. till h. in more detail.

ad. a. process in the posterior fossa.

1. processes outside the brainstem.

The tumors which most frequently occur in this regio are the cerebello-pontine- angle tumor and the tumor cerebelli. The suspicion of the former is the most important indication for the registration of the BAEP, for the BAEP will sometimes show abnormalities before the tumor can be seen on the CT-scan. (Stockard, Stockard and Sharbrough, 1977).

The abnormalities in these cases are not specific. They vary between delayed latency of wave I and prolonged IPL's till absence of one or more waves. Often they are asymmetrical. Sometimes the greatest abnormalities will be found contralateral to the side of the tumor. This, because of compression of the brainstem against the bony structures of the base of the skull.

In the case of tumor cerebelli there are often prolonged IPL's.

Some more references:

Wielaard and Kemp (1979)

Selters and Brackmann (1977)

2. processes inside the brainstem.

In these cases there is often a prolonged CCT to be seen.

One or more waves might be absent. It is not always possible to differentiate the findings from those of an extra-axial process. If the IPL I-III is prolonged, there might be a lesion in the pons; if the IPL III-V is prolonged this is reason to suspect a process in the mesencephalon. If the IPL I-V (CCT) is prolonged one is inclined to suspect a process in pons and mesencephalon.

Some literature:
Starr and Archor (1975)
Starr and Hamilton (1976)
Stockard and Rossiter (1977)

ad. b. Demyelinating diseases.

In case of MS it is possible to find a "silent lesion". For the diagnosis of MS it is necessary to find more than one lesion. Apart from the clinical signs one may find a localisation of a "plaque" by means of the study of EP's. All the modality's of the EP's can be used and in fact it is best to use them all- then the chance of finding a lesion is maximal. Chiappa (1980) found disturbances of the BAEP in 32%, of the SSEP in 60% and of the VEP in 56%. In these cases there is mostly a disturbance of the later components of the BAEP.

Among the many other publications on this subject we mention: Tackmann et al. (1980) and Robinson et al. (1977). In addition there is a publication by Stockard (1976) about the Central Pontine Myelinolysis. In two patients with this condition he found a slowing of conduction in the pontine auditory pathway, which varied in degree with the severity of the clinical manifestations.

ad. c. Comata

The BAEP can be a powerful diagnostic tool when the question arises whether the coma is due to a functional or to an anatomic disturbance of the brainstem. In the former the BAEP will mostly be normal, in the latter there will be abnormalities. So it is also of prognostic value, for in the case of an anatomic lesion the prognosis will be less favourable.

In the case of coma due to severe barbiturate intoxication, the EEG can be isoelectric while the BAEP is still unaffected. Some references:Starr (1975) Uxiel et al. (1982).

ad. d. Intoxications.

Chu et al. (1982) reported an chronic alcohol intoxication. In 66 patients they found 41% abnormal BAEP's indicated by prolonged interwave intervals. The incidence of brainstem abnormalities was related to age, the type and number of alcoholic neurological complications and cerebral atrophy in CT-scan. Acute effects of alcohol were described by Squires et al. (1978) and Jones et al. (1980).

ad. e. Traumata.

In the case of trauma of the nervous system, EP's can provide only a slight improvement on predictions based upon clinical features. The cortical EP's show correlations with these features but not the BAEP (Lindsay et al. 1981). However, there are also more optimistic publications (Tsubokawa et al. 1980). Noseworthy et al. (1981) reported on the postconcussion syndrome. They found in this condition (which is often supposed to be psychogenic significant delays for wave III.

ad. f. Cerebral death.

In case of brain death a flat EEG is recorded. This means that the cerebral cortex is not functioning. However, for more security it is important to be informed about the functioning of the brainstem too. This provides valuable additional information. Starr (1976) reported that in the case of brain death the BAEP is either absent or consists of the presence of just wave I. First wave VII disappears, then VI and so on. It is important to register wave I, for that means that there are no technical or hearing failures.

ad. f. Neuropediatry.

In this discipline it is important to estimate the cerebral maturity in infants with developmental disturbances.

When the child grows older there are remarkable changes of the BAEP: the latencies become shorter and the waves will be better developed. The maturation of the BAEP takes place within the first three years. The best descriptions of this process can be found in Hecox (1974) and Stockard et al. (1981).

ad. h. Neurosurgery.
The application of the BAEP as a monitor in operations of the posterior fossa is described by Hashimoto et al. (1980).

Some rules:

- In cases of serious hearing loss it is necessary to be cautious to conclude neurologic disorders from an abnormal BAEP.
- an abnormality of wave I (for instance in patients with a cerebello-pontine-angle tumor) indicates the exact lateralisation of the lesion. The lateralisation remains doubtful when waves III and/or V are disturbed.
- a tumor of the VIIIth nerve frequently gives asymmetric abnormalities (sometimes predominant on the contralateral side because the brainstem is pushed against the skull). An intra-axial disease of the brainstem usually produces symmetric disturbances.
- in the case of a cerebello-pontine-angle tumor there are no specific patterns.
- a brainstem functional disorder, either caused by intoxication or of metabolic origin, mostly gives a normal BAEP, so a disturbed BAEP points to the presence of an anatomical lesion.
- in supratentorial disorders the BAEP is normal.

8.3 MLAEP

8.3.1 MLAEP in adults

The clinical application of the MLAEP has found acceptance in audiology rather than in neurology or neuropediatrics. Normative data have been collected in adults and to some extent also in children (Table 3.1). The response proved to be a sensitive tool in measuring the auditory threshold. In neuroaudiology the MLAEP may provide a test for lesions in the auditory system central to the brainstem. The MLAEP's appear promising for the diagnosis of acoustic or brainstem tumors, in demyelinating diseases and in central deafness (Wolf, 1978, Greiner, 1978).

8.3.2 MLAEP in infants and children

Engel (1971) identified, in his study of 24 neonates, only 8 MLAEP's and advocated the use of long latency AEP (LLAEP) as a more reliable tool for audiological examinations in infants. Later studies, however, reached opposite conclusions (McRandle 1974, Wolf 1978, Mendel 1977). MLAEP waveforms can be obtained in neonates. The waves resemble those in adults but show considerable asymmetry, in contrast with adults. The most reliable component in neonates is the homolateral NaPaNb component (Wolf 1978). There is no significant difference in latencies as a function of age (Mendel 1977). The amplitudes of the positive waves Po, Pa and Pb increase up to the age of three to four and tend to decline in adulthood (Mendelson).

In conclusion, we may consider the MLAEP as a potentially reliable tool in the diagnostic armamentarium for the determination of audiologic as well as neuroaudiologic disorders in neonates, young children and adults.

8.4 AEP

For the clinical use of the cortical AEP two different fields find its application useful. One is in audiology where the cortical ERA is used with different techniques and on which the opinions are rather different with respect to its value; and in neurology as an aid in the investigation of central lesions located above the lower brainstem. Both areas of application are still rather modest. One of the causes is undoubtly the emphasis which has been placed on the visual evoked potential as a powerful tool in clinical diagnosis which has diminished research in the clinical usefulness of the AEP somewhat. In other fields, especially in research concerning attention, the AEP is frequently used.

The AEP is somewhat underrated as a clinical tool although stimulus and response are much simpler than for the VEP and SSEP. It is to be expected that with better knowledge of and more experience with the AEP its use in the clinical neurophysiology laboratory will increase. For investigation of the pathology of the periferal auditory system the BAEP is a very powerful diagnostic method.

8.4.1 ERA and AEP

Since the mid-fifties cortical ERA has had a long history of widely differing appraisals. Rapin et al. (1977) concluded that as a screening method for the assessment of the hearing threshold in children it is not sufficient and that electrocochleography or BAEP are better choices. They recommend the use of the cortical ERA in the later stage,when periferal causes for hearing disturbances are excluded. Campbell et al. (1977), on the contrary, find that the use of cortical ERA using clicktrains with repetition rates of 12-16 per second are a simple, reliable tool for the objective assessment of hearing thresholds within ± 5 dB.

Others using FM-stimuli also find reliable threshold determinations to be possible. It is of importance that the clinical neurophysiologist using the technique has sufficient experience as it has been demonstrated that false-positives and false-negatives decline when the laboratory gains experience. Most of the authors publishing series of ERA results point at the necessity of computerized peak detection and identification to increase reliability (e.g. Rapin et al. 1977).

The most important diagnostic help is with autistic and non-cooperative children, young children (under one year) and disturbed adults. Both cortical AEP and BAEP are preferred over cochleography as this is an invasive technique. Whenever possible both ERA using cortical and brainstem response is to be used as their information is complementary.

8.4.2 Neurological applications

For neurology, application of the AEP as an aid to diagnostic problems is mainly centered on leasions in the parieto temporal area. Tumors and vascular disorders display distinct asymmetries in latency and amplitudes both for tone stimuli and for clicks. An important aspect is that in righthanded patients an abnormal or even absent AEP in the left hemisphere, indicative of disfunctioning of the underlying hemisphere is usually not perceived by the patient (Perronet et Michel 1977). Right hemispheric dysfunctions usually give rise to language disturbances. Generally when cortical parietotemporal lesions are suspected the AEP is often abnormal. It may be usefull to differ between frontal and temporal dysfunctions as the former often don't change the AEP. In various cortical hearing and language dysfunctions the AEP can be used to differentiate between an organic or psychic origin of the symptoms. Martineau and coworkers (1980) have used the AEP with other EP's to differentiate between mentally retarded and autistic children. In psychiatric patients the AEP may be helpful as it can be used as an indicator of the amelioration or deterioration of the patients mental condition (Roth 1980).
Up till now, the FM-AEP have mainly been used for periferal lesions in the auditory pathway (Arlinger 1976). More experience has to be gained to evaluate its clinical use in early detection of lesions in the higher parts of the auditory system.

9. EXAMPLES OF BAEP IN DAILY PRACTICE.

a. Normal BAEP (Fig. 3.28).

b. Perception deafness left (Fig. 3.29).

c. Total deafness right (Fig. 3.30).

d. Deafness left after lues infection (Fig. 3.31).

e. Lues cerebro-spinalis (Fig. 3.32).

f. Cerebello-pontine-angle tumor right (Fig. 3.33).

g. Acoustic neuroma right (Fig. 3.34).

h. Cerebello-pontine-angle tumor right (Fig. 3.35).

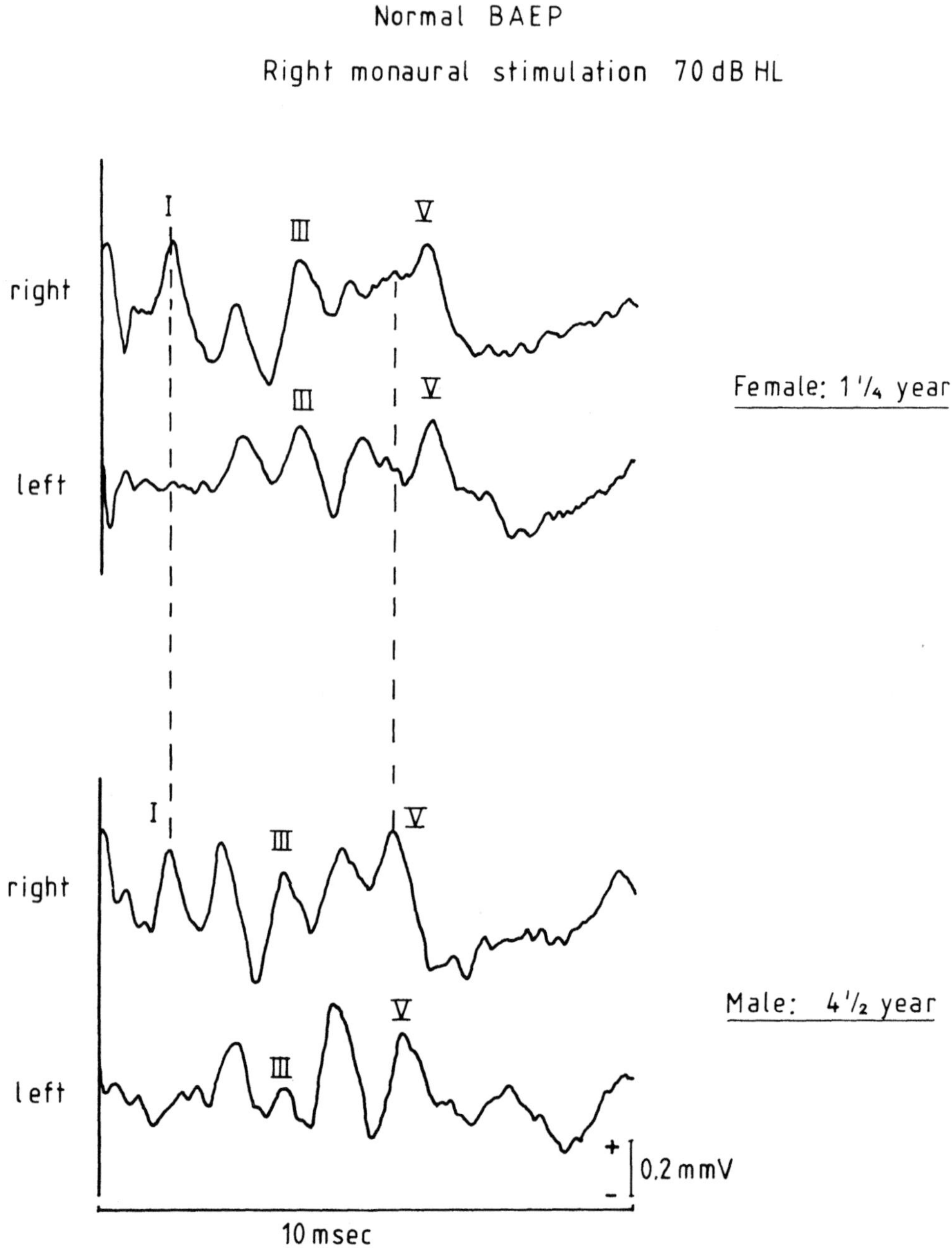
Normal BAEP
Right monaural stimulation 70 dB HL
I
III
V
right
left
Female: 1 1/4 year
I
III
V
right
left
Male: 4 1/2 year
+
0.2 mmV
-
10 msec

Fig. 3.28

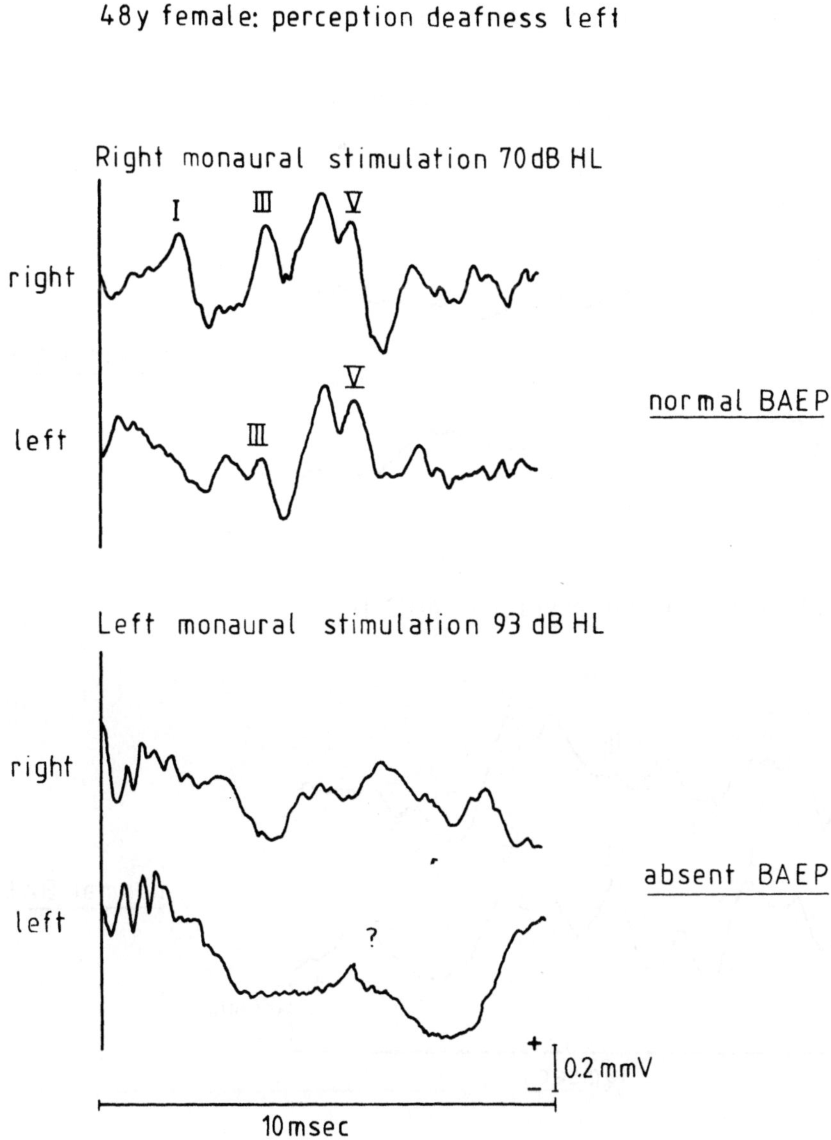

48y female: perception deafness left
Right monaural stimulation 70dB HL
I
III
V
right
left
III
V
normal BAEP
Left monaural stimulation 93 dB HL
right
left
?
absent BAEP
+
0.2 mmV
-
10msec

Fig. 3.29

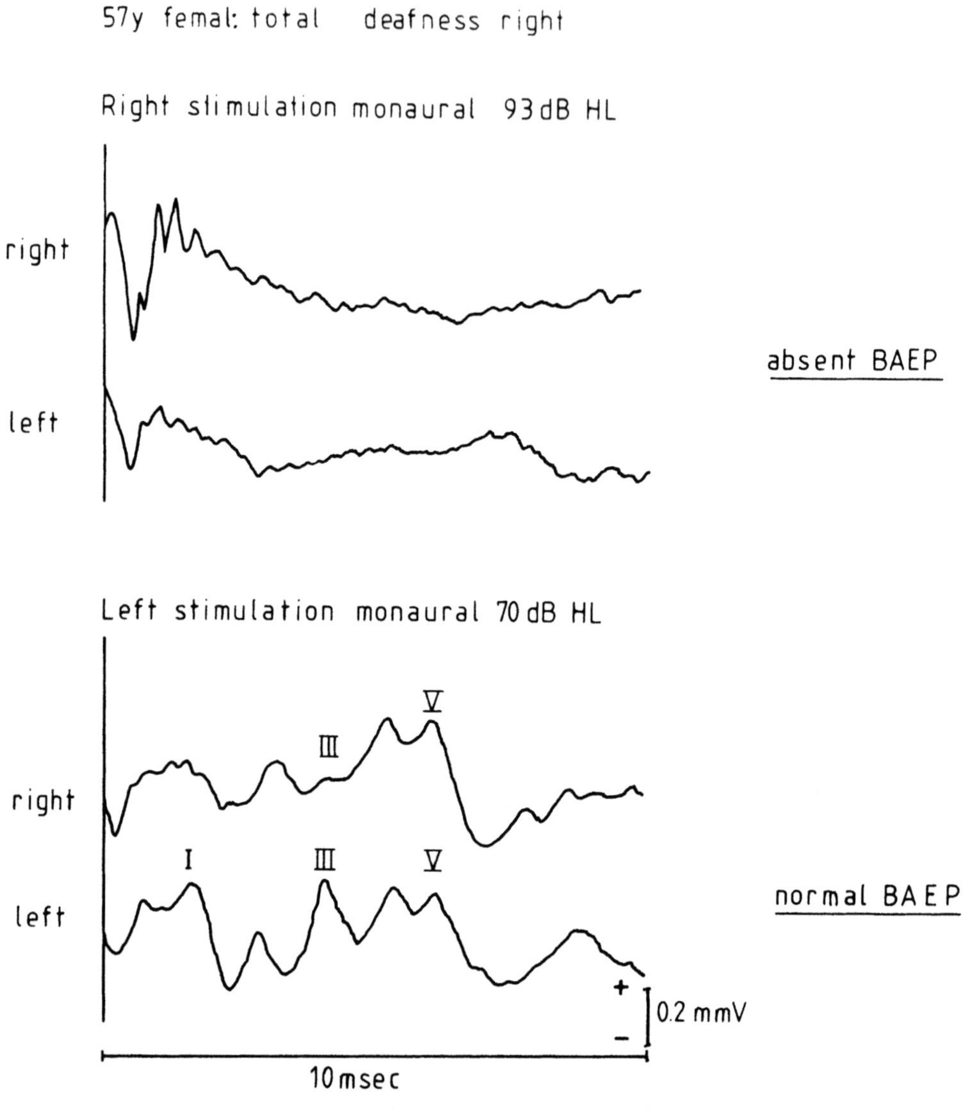
57y femal: total deafness right
Right stimulation monaural 93dB HL
right
left
absent BAEP
Left stimulation monaural 70dB HL
V
III
right
I
III
V
left
normal BAEP
+
0.2 mmV
−
10 msec

Fig. 3.30

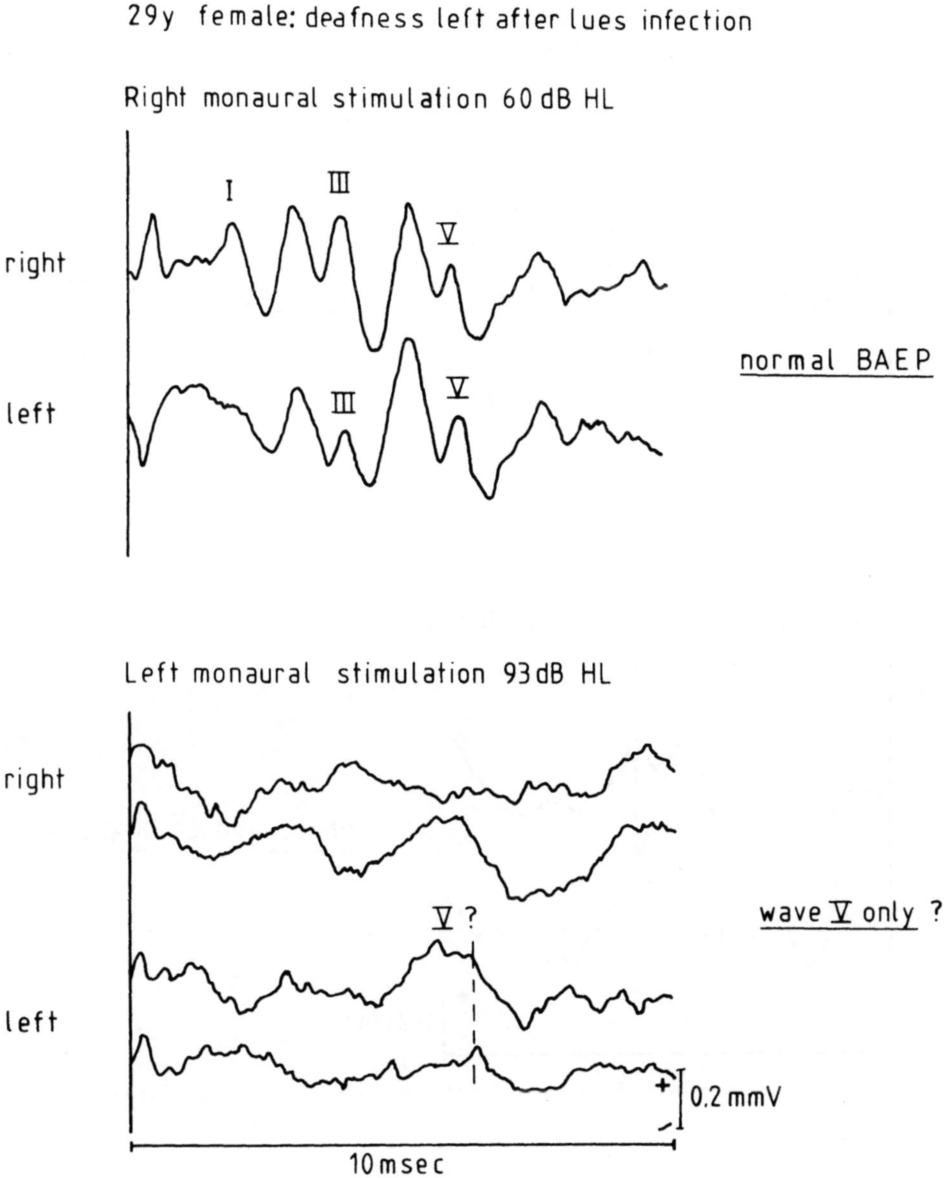
29y female: deafness left after lues infection
Right monaural stimulation 60 dB HL
I
III
V
right
III
V
left
normal BAEP
Left monaural stimulation 93dB HL
right
V ?
wave V only ?
left
+
0.2 mmV
-
10 msec

Fig. 3.31

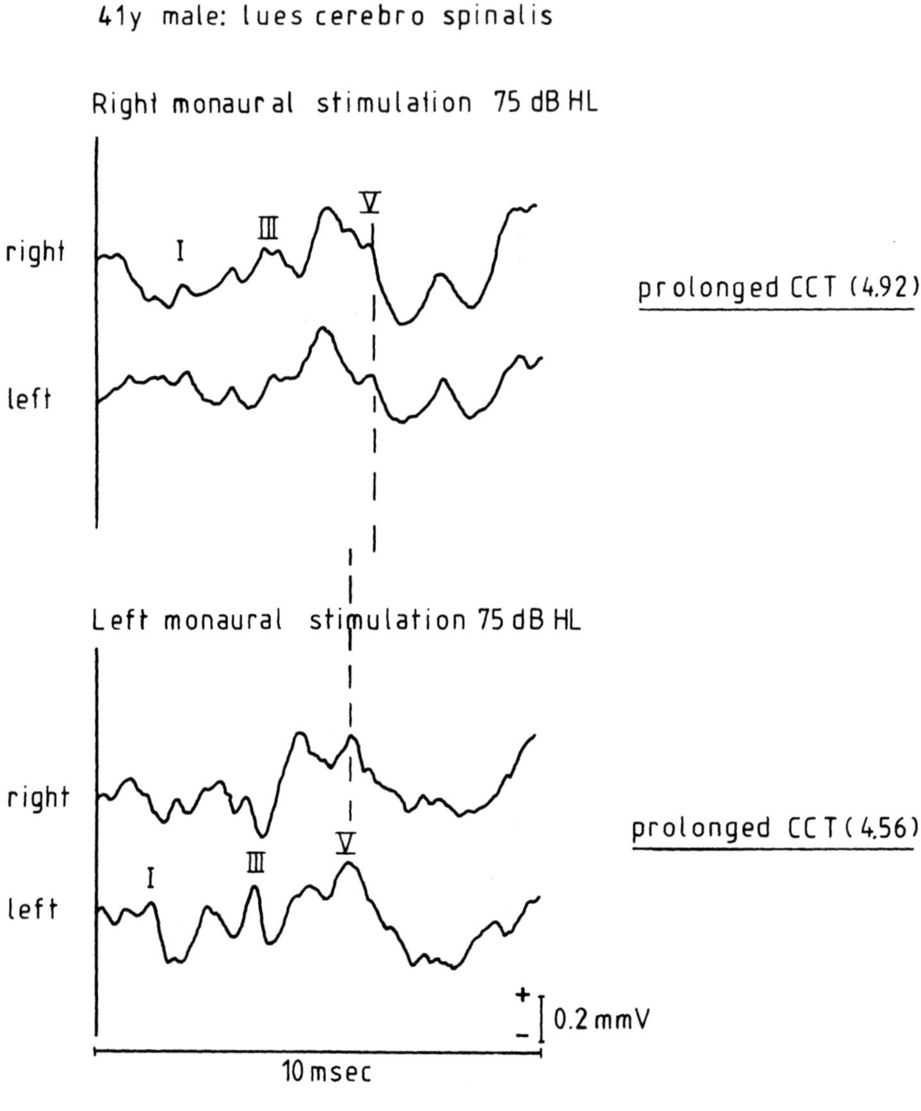
41y male: lues cerebro spinalis
Right monaural stimulation 75 dB HL
right
left
I
III
V
prolonged CCT (4.92)
Left monaural stimulation 75 dB HL
right
left
I
III
V
prolonged CCT (4.56)
+
-
0.2 mmV
10 msec

Fig. 3.32

39 y male: cerebello-pontine-angle tumor (schwannoma) right.

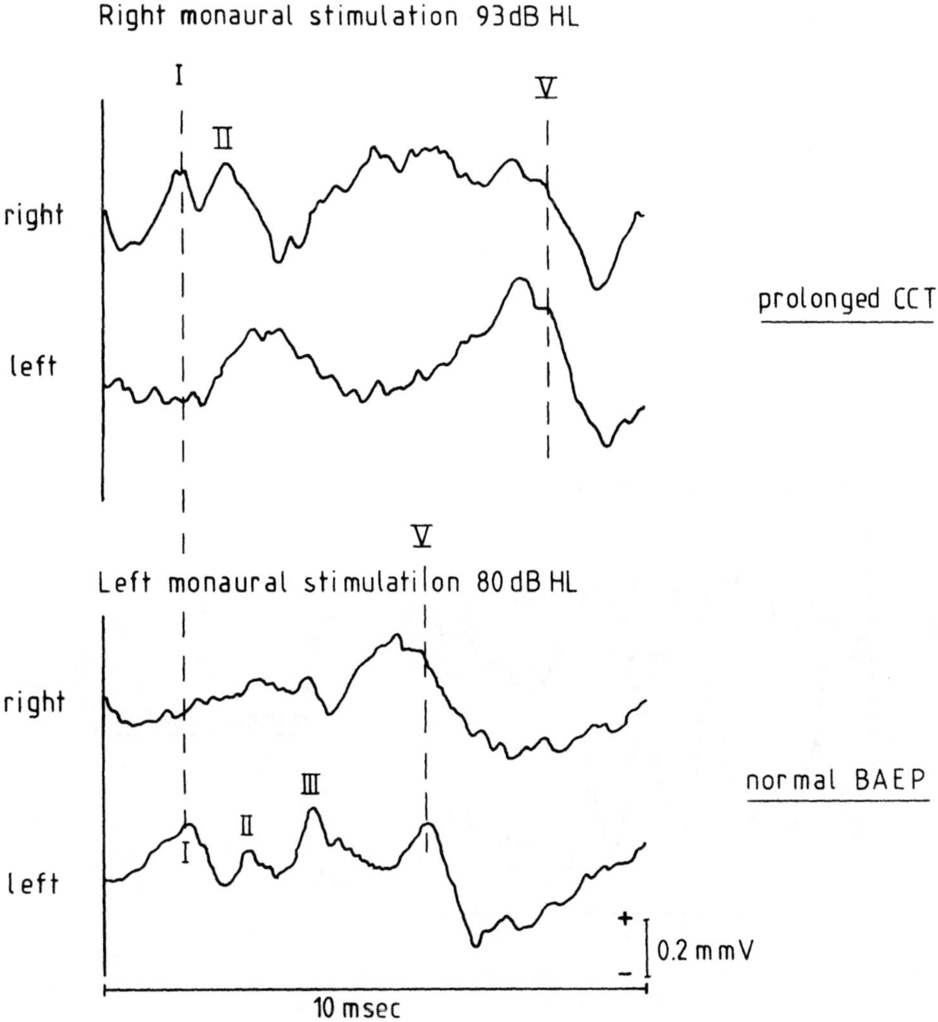

Fig. 3.33

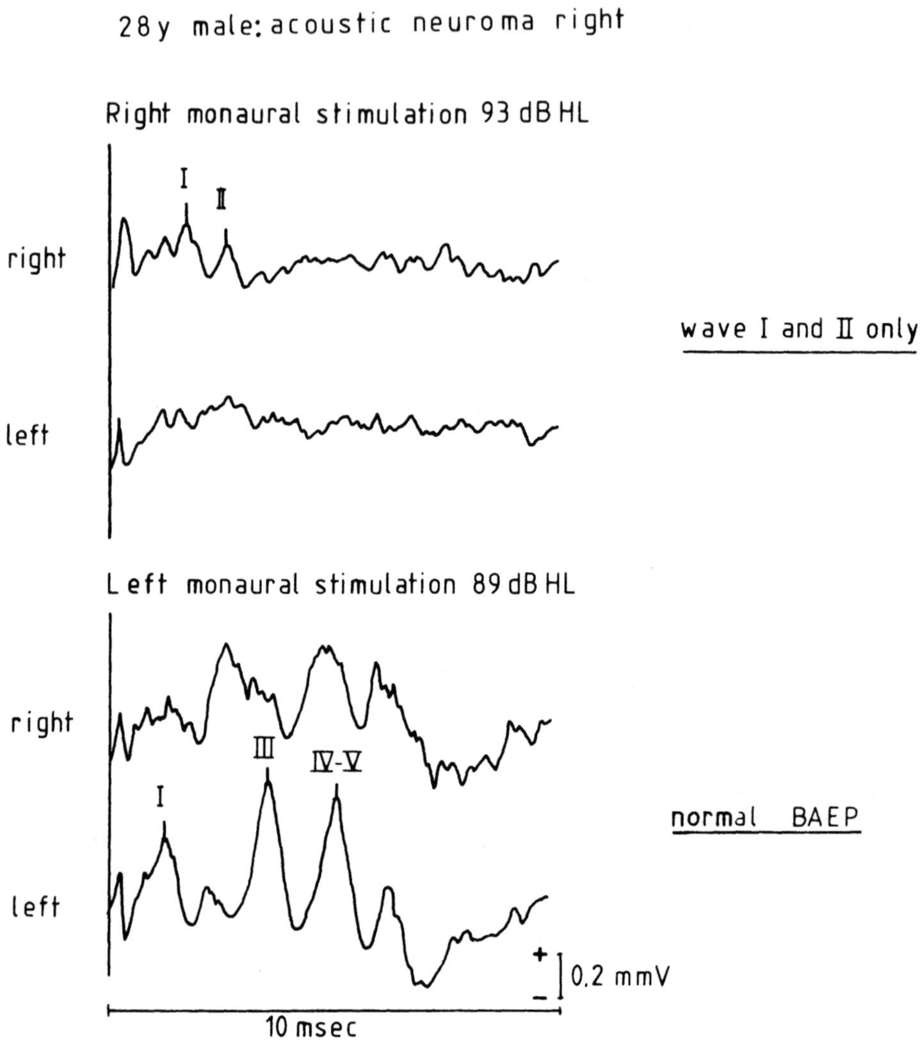

28 y male: acoustic neuroma right
Right monaural stimulation 93 dB HL
I
II
right
left
wave I and II only
Left monaural stimulation 89 dB HL
right
III
IV-V
I
left
normal BAEP
+
0.2 mmV
–
10 msec

Fig. 3.34

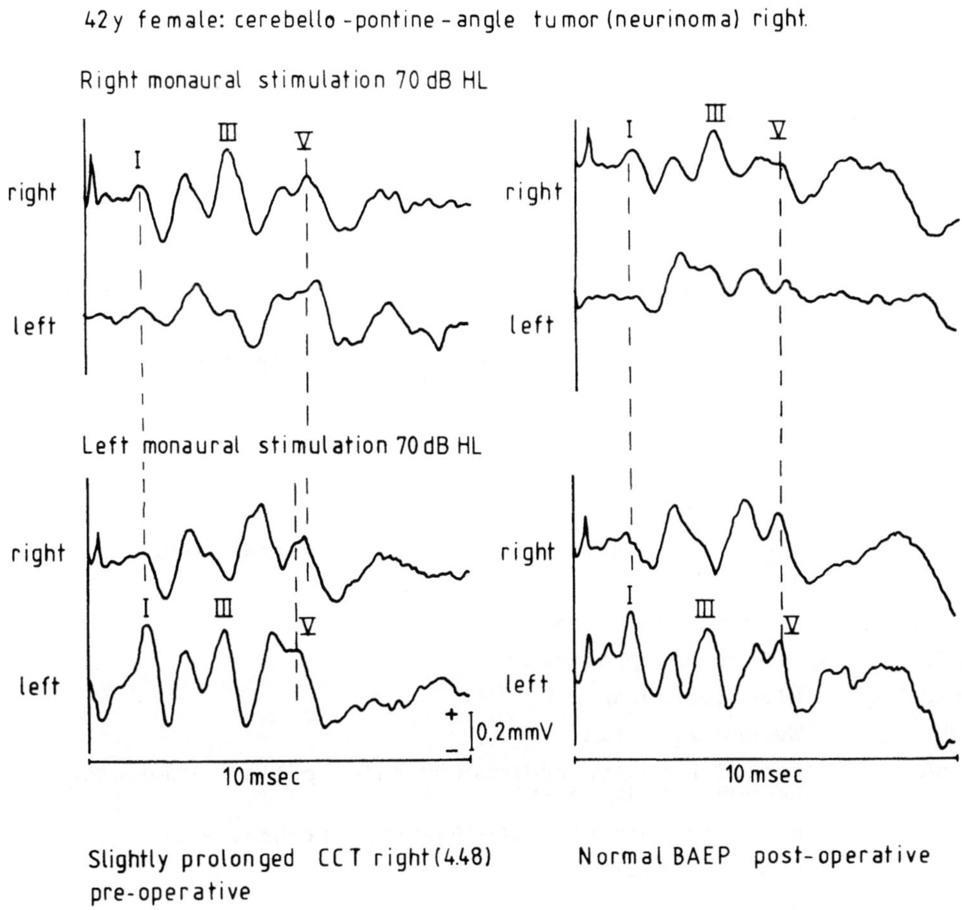
42y female: cerebello-pontine-angle tumor (neurinoma) right.
Right monaural stimulation 70 dB HL
I
III
V
right
left
Left monaural stimulation 70 dB HL
right
left
+
0.2mmV
-
10 msec
Slightly prolonged CCT right (4.48)
pre-operative
Normal BAEP post-operative

Fig. 3.35

10. ABBREVIATIONS

Acontro	electrode position 10-20 system - ear contralateral
AEP	Auditory Evoked Potential
Aipsi	electrode position 10-20 system - ear ipsilateral
AP	Action Potential
BAEP	Brainstem Auditory Evoked Potential
CCT	Central Conduction Time
CM	Cochlear Microphonics
Cz	electrode position 10-20 system central zero
dB	decibel
ECochG	Electrocochleogram
EEG	Electroencephalogram
EMG	Electromyogram
EP	Evoked Potential
ERA	Evoked Response Audiometry (also called: Electric Response Audiometry)
Fz	electrode position 10-20 system frontal zero
HL	Hearing level
Hz	Hertz
i.m.	intra-muscular
IPL	Inter Peak Latency
MLAEP	Middle Latency Auditive Evoked Potential
mmV	milli-milliVolt= microVolt
ms	millisecond
mV	milliVolt
PL	Peak Latency
SD	Standard Deviation
SL	Sensation Level
SPL	Sound Pressure Level
$\sqrt{n}$	square root of the number (n=number)
10^4	10.000

REFFERENCES AND RECOMMENDED LITERATURE

Ainslie, P.J. and Boston, J.R.: Comparison of brainstem auditory evoked potentials for monaural and binaural stimuli. Electroencephal. and Clin. Neurophysiol. 49: 291-302, (1980).

Allen, A.R. and Starr, A.: Auditory brainstem potentials in monkey (M. Mulatta) and man. Electroencephal. and Clin. Neurophysiology, 45: 53-63, (1978).

Aminoff, M.M. (Ed.): Electrophysiologic Approaches to Neurologic Diagnosis. Churchill Livingstone, New York, (in press), 541-567, (1979).

Arlinger, S.: Auditory responses to frequency ramps; a psycho-acoustic and electrophysiological study. Linköping University,(Linköping University Medical Dissertations, no. 40) (1976).

Baltimore, MD, Williams, Wilkins,1977: Handbook of physiology: Section I: the nervous system (formerly Neurophysiology). Vol. I. Cellular biology of neurons; Ed. by E.R. Kandel, (1977).

Bickford, R.G., Jacobson, J.L., Cody, D.T.R.: Nature of averaged evoked potentials to sound and other stimuli in man. Ann. N.Y. Acad. Sci., 112, 204-223, (1964).

Blom, J.L. and van Blokland, A.W.: Analysis Technique of evoked responses and peak identification for use in mental load research. In: Matujeck, M. and G.K. Schenk, (Eds.) The quantification of the electroencephalogram II; proceedings of the 2nd Symposium of the Study Group for EEG-technology, Jogny sur Vevey (may 1975).

Blom, J.L., Barth, P.G. , Visser, S.L.: The visual evoked response in the first six years of life. Electroencephal. Clin. Neurophysiol. 48, 395-405, (1980).

Bostock, H. and Jarvis, J.: Changes in the form of cerebral evoked responses related to the speed of simple reaction-time. Electroencephal. Clin. Neurophysiol., 29, 137-145, (1970).

Brodal, A.: Neurological Anatomy in relation to clinical medicine. Oxford University Press London, (1979).

Buchwald, J.S., Huang, C.M.: Far Field Aucoustic Response: Origins in the cat. Science, Vol. (1975).

Campbell, F.W., Atkinson, J., Francis, M.R. and Green, D.M.: Estimation of auditory thresholds using evoked potentials. A clinical screening test.
In: Desmedt (Ed.). Auditory Evoked Potentials in man. Progr. in Clin. Neurophysiol. vol. 2, 68-79, (Karger, Basel, 1977).

Carpenter, M.B.: Human Neuroanatomy,7th ed. The Williams and Williams Company Baltimore, (1976).

Celesia, G.G.: Auditory evoked responses: intracranial and extracranial average evoked responses. Arch. Neurol. 19, 430-437, (1968).

Chiappa, M.D., Keith, H.: Pattern shift visual, brainstem auditory and shortlatency somatosensory evoked potentials in multiple sclerosis. Neurology 30, (2), 110-123, (1980).

Chu, N.S., Squires, K.H. and Starr, A.: Auditory brainstem responses in chronic alcoholic patients. Electroencephal. and Clin. Neurophysiol., 54, 418-425, (1982).

Church, M.W. Williams, H.L.: Dose and time dependent effects of ethanol on brainstem auditory evoked responses in young adult males. Electroencephal. and Clin. Neurophysiol., 54, 161-174, (1982).

Cohen, M.M.: Coronal Topography of the MLAEP's in man. Electroencephal. and Clin. Neurophysiol., 53, 231-236, (1982).

Davis, H., Hirsh, S.K.: A slow brainstem response for low-frequency audiometry. Audiology, 18, 445-461, (1979).

Davis, H. and Yoshie, N.: Human evoked cortical responses to auditory stimuli. Physiologist, 6, 164, (1963).

Don, M. and Eggermont, J.J.: Analysis of the click-evoked brainstem potentials in man using high-pass noise masking. The journal of the acoustical society of America, Vol. 63, (4), (1978).

Edwards, M.R., Buchwald, J.S., Tanguay, P.E., Schwafd, A.J.: Sources of variability in auditory brainstem evoked potential measures over time. Electroencephal. and Clin. Neurophysiol. 53, 125-132, (1982).

Eggermont, J.J., Don, M., Brachmann: Electrocochleography and auditory brainstem electro responses in patients with pontine angle tumors. The annals of otology, rhinology and laryngology, Supp. 75, Vol. 89, no. 6, part 2. (nov./dec. 1980).

Fujita, M., Hosoki, M., Miyazaki, M.: Brainstem Evoked Responses in Spinocerebellar Degeneration and Wilson Disease.

Funasaka, S., Kodera, K. and Abe, H.: The auditory stimuli to evoke a clear average response at behavioral threshold. Audiology 13, 162-172 (1974).

Galambos, R.M.D. and Hecox, K.E.: Clinical Applications of the Auditory Brainstem Response. Otolaryngologic Clinics of North America, Vol. 11 no. 3, (figure 4 and 5 are exchanged) (1978).

Galambos, R., Makeig, S., Talmaohoff, P.J.: A 40-Hz auditory potential recorded from the human scalp. Proc. Natl. Acad. Sci., 78, 4, 2643-2747, (19

Geisler, C.D., Frishkopf, L.S., Rosenblith, W.A.: Extracranial responses to acoustic clicks in man. Science, 128, 1210-1211, (1958).

Gerin, P., Pernier, J., Peronnet, F.: Amplitude des potentiels évoqués noyens auditifs du vertex et intensité des stimuli. Brain Research 36, 89-100, (1972).

Goff, G.D., Matumiya, Y., Allison, T., Goff, W.R.: The scalp topography of human somatosensory and auditory evoked potentials. Electroencephal. Clin. Neurophysiol. 42, 57-76, (1977).

Greiner, G.F., Conraux, C., Fehlet, P.: Audiométrie objective par enregistrement des potentiels évoqués auditifs précoces et semi-précoces par voie externe. Ann. Otolaryng. (Paris) 95, 33-41, (1978).

Harker, A., Hosick, E., Voots, R.J. and Mendel, I.M.: Influence of succinylcholine on middle component auditory evoked potentials. Arch. Otolaryngol. 103, 133-137, (1977).

Hashimoto, I.: Auditory evoked potentials from the human midbrain: slow brainstem responses. Electroencephal. Clin. Neurophysiol. 53, 652-657, (1982).

Hashimoto, I., Ishiyama, Y., Tozuka, G.: Bilaterally Recorded Brain Stem Auditory Evoked Responses. Arch. Neurol. Vol. 36, (1979).

Hecox, K., Galambos, R., La Jolla, Ph.D.: Brainstem Auditory Evoked Responses in Human Infants and Adults. Arch. Otolaryngol. Vol 99. (Jan. 1974).

Hecox, K.E.: Brainstem Evoked Potentials - Neurological Applications. Center for health sciences university of Wisconsin-Madison, (1980).

Hughes, J.R., Fino, J. Gagnon, L.: The importance of phase of stimulus and the reference recording electrode in brainstem auditory evoked potentials. Electroencephal. and Clin. Neurophysiol. 51, 611-623, (1981).

Jewett, D.L.: Volume-conducted potentials in response to auditory stimuli as detected by averaging in the cat. Electroencephal. and Clin. Neurophysiol., 28, 609-618, (1970).

Jewett, D.L., Romano, M.N. and Williston, J.S.: Human auditory evoked potentials: possible brainstem components detected out the scalp. Science 167, 1517-1518, (1970).

Jewett, D.L. and Williston, J.S.: Auditory evoked far fields averaged from the scalp of humans. Brain, 94, 681-696, (1971).

John, E.R., Baird, H., Fridman, J., Bergelsono, M.: Normative values for brainstem auditory evoked potentials obtained by digital filtering and automatic peak detection. Electroencephal. and Clin. Neurophysiol., 54, 153-160, (1982

Jones, L. Komsmoghn, S., Smith, E., Herding, C.: Brainstem auditory evoked potentials recorded in patients with confirmed and clinically suspected brainstem disorders. Journal of Electrophysiology Technology, 8, 135-150, (1982).

Jones, T.A., Stockard, J.J. and Weidner, W.J.: The effects of temperature and acute alcohol intoxication on brainstem auditory evoked potentials in the cat. Electroencephal. and Clin. Neurophysiol. , 49, 23-30, (1980).

Kaga, K., Hink, R.F., Shinoda, Y., Suzuki, J.: Evidence for a primary cortical origin of a middle latency auditory evoked potential in cats. Electroencephal. and Clin. Neurophysiol. 50, 254-266, (1980).

Kogatal, S., Rosenberg, C., Rudd, D., Dunkle, M., Horenstein, S.: Auditory Evoked Potentials in bacterial Meningitis.Arch. Neurol. Vol. 38, (nov 1981).

Kraus, N., Özdamar, O., Hier, D., Stein, L.: Auditory middle latency responses (MLR's) in patients with cortical lesions. Electroencephal. and Clin. Neurophysiol., 54, 275-287, (1982).

Lindsay, K.W., Carlin, J., Kennedy, I., Maines, A., Teasdale, C.M.: Evoked Potentials in severe head injury - analysis and relation to outcome. Journal of Neurol. Neurosurgery and psych., 44, 796-802, (1981).

Lolas, F., Hoeppner, T.J. and Morrell, F.: The BAER profile: A method for reporting electrophysiologic brainstem examinations in clinical practice. Neurology, 29, 242-244, (1979).

Markand, O.N., Garg, B.P., Demeyer, W.E., Warner, C., Worth, R.M.: Brainstem Auditory Visual and Somatosensory evoked potentials in leukodystrophies. Electroencephal. and Clin. Neurophysiol. 54, 39-48, (1982).

Martineau, J., Laffont, F., Brunneau, N., Roux, S. and Lelord, G.: Event -related potentials evoked by sensory stimulation in normal, mentaly retarded and autistic children. Electroencephal Clin. Neurophysiol. 48, 140-153, (1980).

Mast, T.E.: Short latency human evoked responses to clicks. J. Appl. Physiol. 20, 725-730, (1965).

McFarland, W.H. and Vivion, M.C.: Reexamination of effects of stimulus rate and number on the middle components of the averaged evoked response. Audiology, 14, 456 (1975).

McFarland, W.H., Vivion, M.C., Goldstein, R.: Middle components of the AER to tone-pips in normal-hearing and hearing-impaired subjects. J. Speech Hearing, Res. 20, 781-798, (1977).

McRandle, C.C., Smith, M.A., Goldstein, R.: Early averaged electro-encephalic responses to clicks in neonates. Ann. Otol. Rhinol. Laryngol. 83, 695-702, (1974).

Mendel, I.M., Goldstein, R.: The effect of test conditions on the early components of the averaged electro-encephalic response. J. Speech Hearing Res.12, 344-350, (1969).

Mendel, M.I.: Influence of stimulus level and sleep stage on the early components of the averaged electro-encephalic response to clicks during all night sleep. J. Speech Hearing, Res. 17, 5-17, (1974).

Mendel, M.I., Hesick, E.C., Windman, T.R., Davis, H., Hirsh, S.K., Dinges, D.F.D.: Audiometric comparison of the middle and late component of the adult auditory evoked potentials awake and asleep. Electroencephal. and Clin. Neurophysiol., 38, 27-33, (1975).

Mendel, M.I., Hosick, E.C.: Effects of secobarbital on the early components of the auditory evoked potentials. Revue de Laryngologie 96, 178-184, (1975).

Mendel, M.I., Adkinson, C.D., Harker, L.A.: Middle components of the auditory evoked potentials in infants. Ann. Otol. Rhinol. Laryngol. 86, 1-7, (1977).

Monnier, M.: Function of the Nervous System . Vol. 3 Sensory Functions and Perception. Elsevier, Amsterdam,London, New York, (1975).

Musiek, F.E., Geurkink, N.A.: Auditory brainstem and middle latency evoked response sensitivity near threshold. Ann. Otol. Rhinol. Laryngol. 90, 236-240, (1981).

Noseworthy, J.H., Miller, J., Murray, T., Regan, D.: Auditory Brainstem Responses in Postconcussion syndrome. Arch. of Neurol. Vol. 38, (May 1981).

Olphen van, A.F., Rodenburg, M. and Verwey, C.: Distribution of Brainstem Responses to Acoustic Stimuli over the Human Scalp. Audiology 17, 511-518, (1978).

Olphen van, A.F., Rodenburg, M. and Verwey, C.: Influence of the Stimulus Repetition Rate on Brain-Stem-Evoked responses in Man. Audiology 18, 388-394, (1979).

Parring, A., Salomon, G., Elberling, C., Larsen, B., Lassen, N.A.: Middle components of the auditory evoked response in bilateral temporal lobe lesions. Scand. Audiol. 9, 161-167, (1980).

Peronet, F. and Michel, F.: The asymmetry of the auditory evoked potentials with brain lesies. In: Auditory evoked potentials in man: psychopharmacology correlates of EP's,editor: J.E. Desmedt, Basel, New York, (Karger, 1977).

Picton, T.W., Hillyard, S.A., Krausz, S.I. and Galambos, R.: Human Auditory evoked potentials. I: evaluation of components. Electroencephal. and Clin. Neurophysiol., 36, 179-190, (1974).

Picton, T.W. and Hink, R.F.: Evoked potentials: How? What? and Why? American Journal of EEG Technology, Vol. 14, (1974).

Picton, T.W., Woods, D.L. and Proulx, G.B.: Human Auditory sustained Potentials. II. Stimulus Relationships. Electro-encephal. and Clin. Neurophysiol., 45, 198-210, (1978).

Praher, D.K. and Gibson, W.P.R.: Brainstem auditory evoked potential significant latency differences between ipsilateral and contralateral stimulation. Electroencephal. and Clin. Neurophysiol. 50, 240-246, (1980).

Pratt, H., Bleich, N.: Auditory Brainstem Potentials Evoked by clicks in Notch-Filtered Masking Noise. Electroencephal. and Clin. Neurophysiol. 53, 417-426, (1982).

Rapin, I. and Schimmel, H.: Assessment of auditory sensitivity in infants and in un-cooperative handicapped children by using the late components of the average auditory evoked Potential. In: Auditory evoked potentials in man: Psycho-pharmacology, correlates of EP's, ed. J.E. Desmedt, Basel, New York, (Karger, 1977).

Robinson, K. and Rudge, P.: Abnormalities of the auditory evoked potentials in patients with multiple sclerosis. Brain, 100, 19-40, (1977).

Rose, D.E., Keating, L.W., Hedgecock, L.D., Scheurs, K.H. and Miller, K.E.: Aspects of acoustically evoked responses. Arch. Otolaryng. 94, 347-350, (1978).

Rosenhall, U., Rompe, G.: Auditory brainstem response in syphilis. B.J. Venen Dis. 57, 241-245, (1981).

Rossini, P.M., Cambi, D., Marchionno, L., David, P., Sollazzo, D.: Cephalic and Non cephalic References in Brainstem Evoked Potential Recording. App. Neurophysiol. 43, 313-322, (1980).

Rossini, P.M., Kula, R.W., House, W.J., Cracco, R.Q.: Alteration of Brainstem Auditory Evoked Responses following cardio-respiratory arrest and resuscitation. Electroencephal. and Clin. Neurophysiol., 54, 323-334, (1982).

Roth, W.T., Horvath, T.B., Pfefferbaum, A. and Kopell, B.S.: Event related potentials in schizophrenics. Electroencephal. and Clin. Neurophysiol. 48, 127-139, (1980).

Rowe, M.J.: Normale variability of the brainstem auditory evoked response in young and old adult subjects. Electroencephal. and Clin. Neurophysiol., 44, 459-470, (1978).

Rowe, M.J., Carlson, Ch.: Brainstem Auditory Evoked Potentials in postcontussion Dizziness. Arch. Neurol. Vol. 37, (nov 1980).

Schulman-Galambos, C. and Galambos, R.: Brainstem Auditory- evoked responses in premature infants. Journal of Speech and Hearing Research 18, 456-465, (1975).

Schulman-Galambos, C. and Galambos, D.: Recording the brainstem Evoked Response. Preliminary draft, (1975).

Selters, W., Brachmann, D.E.: Acoustic Tumor Detection With Brain Stem Electric Response Audiometry, Arch. Otolaryngology, Vol. 103, (april 1977).

Sherman, J.: Advanced Diagnostic Procedures. The Nicolet CA-1000-Series 2 no. 11, (1977).

Sohmer, H., Feinmesser, M. and Szabo, G.: Sources of electro-cochleographic responses as studied in patients with brain damage. Electroencephal. and Clin. Neurophysiol. 37, 663-669, (1974).

Squires, N., Aine, C., Buchwald, J., Norman, R. and Galbraith, G.: Auditory brainstem response abnormalities in severely and profoundly retarded adults. Electroencephal. and Clin. Neurophysiol., 50, 172-185, (1980).

Squires, K.C., Chu, N.S., Starr, A.: Acute effects of alcohol on auditory brainstem potentials in humans. Science, vol. 201, (1978).

Starr, A.: Auditory brainstem responses in brain death. Brain 99, 543-554, (1976).

Starr, A., Achor, L.J.: Auditory brainstem responses in neurological disease. Arch. Neurol. Vol. 32, (1975).

Starr, A. and Hamilton, A.E.: Correlation between confirmed sites of neurological lesions and abnormalities of far-field auditory brainstem responses. Electroencephal. and Clin. Neurophysiol.,41, 595-608, (1976).

Starr, A., Amlie, R.N., Martin, W.H. and Sanders, S.: Development of auditory function in newborn infants revealed by auditory brainstem potentials. Pediatrics, 60, 831-839, (1977).

Stockard, J.J., Rossiter, V.S., Widerholt, W.C. and Kobayashi, R.M.: Brainstem Auditory evoked responses in suspected central pontine myelinolysis. Arch. Neurol. Vol. 33, (1976).

Stockard, J.A. and Westmoreland, F.B.: Technical Consideration in the recording and interpretation of the Brainstem Auditory Evoked Potential for neonatal Neurologic Diagnosis. A.M. J. EEG Technol., 21, 31-54, (1981).

Stockard, J.E., Stockard, J.J., Westmoreland, B.F. and Corfits, J.L.: Normal variation as a function of stimulus and subject characteristics. Arch. Neurol. Vol. 36, (1979).

Stockard, J.J. and Rossiter, V.S.: Clinical and pathologic correlates of brainstem auditory response abnormalities. Neurology, Vol. 27, no. 4 pp. 316-325, (1977).

Stockard, J.J., Stockard, J.E., Sharbrough, F.W.: Detection and localization of occult lesions with brainstem auditory responses. Mayo Clin. Proc. 52, 761-769, (1977).

Stockard, J.J., Stockard, J.E. and Sharbrough, F.W.: Nonpathologi factors influencing brainstem auditory evoked potentials. Am. J. EEG Technol. 18, 177- 209, (1978).

Stürzebecher, E., Warnecke, K.D., Wagner, H.: ERA-kennlinien bei normal hörenden Personen. Actua Otolaryngol. 80, 294-300, (1975).

Tackmann, W., Strenge, H., Barth, R., Sojha-Raytschelf, A.: Auditory Brainstem Evoked Potentials in patients with Multiple Sclerosis, Enc. Neurol. 19, 396-401, (1980).

Thornton, A.R., Mendel, M.I., Anderson, C.V.A.: Effects of stimulus frequency and intensity of the middle components of the averaged auditory electroencephalographic response. J. Speech Hearing, Res. 20, 81-94, (1977).

Tsubokawa, T., Nishimoto, H. Yamamoto, T., Kitamura, M., Katayama, Y. Moriyasu, N.: Assessment of brainstem damage by the auditory brainstem response in acute severe head injury. Journal Neurol. Neurosurg. and Psych. 43, 1005-1011, (1980).

Vaughan, H.G. and Ritter, W.: The sources of auditory evoked responses recorded from the human scalp. Electroencephal. and Clin. Neurophysiol. 28, 360-367, (1970).

Vivion, M.C.: Clinical status of evoked response audiometry. The Laryngoscope 90, 437-447, (1980).

Vivion, M.C., Goldstein, R., Wolf, K.E., McFarland, W.H.: Middle components of human auditory averaged electro-encephalic response: waveform variations during averaging. Audiology 16, 21-37, (1970).

Wielaard, B. and Kemp, B.: Auditory brainstem evoked responses in brainstem compression due to posterior fossa tumors. Neurol. Neurosurg. Vol. 81-3, (1979).

Wolf, K.E., Goldstein, R.: Middle component averaged electro-encephalic responses to tonal stimuli from normal neonates. Arch. Otolaryngol. 104, 508-513, (1978).

Wrege, K.S., Starr, A.: Binaural Interaction in Human Auditory Brainstem Evoked Potentials. Arch. Neurol. Vol 38, (Sept. 1981).

Zerlin, S., Maunton, R.F., Mowry, H.J.: The early evoked cortical response to third-ectace clicks and tones. Audiology 12, 242-249, (1973).

Books:

Auditory Evoked Potentials in Man.; Progress in Clinical Neurophysiology, Vol. 2. Editor: Desmedt, J.E. (S.Karger Basel).

EEG Technology.: Cooper, R., Osselton, J.W., Shaw, J.C. (Butterworths, London, 1974).

Essentials of clinical electric response audiometry.: Gibson W.P.R. (Churchill Livingstone, Edinburgh, 1978).

Evoked Potentials in Clinical Testing: Edit. by A.M. Halliday, (Churchill Livingstone, Edinburgh, 1982).

Evozierte Potentiale: M. Stöhr et al. (Springer-Verlag, Berlin, 1982).

Principles of Neural Science.: Kandel, E.R., Schwartz, J.H., Elsevier, North Holland (1981).

Chapter IV

Visual Evoked Potentials

S.L. Visser
G.H.M. van Lith
J.M. Thijssen

1. INTRODUCTION

Of the cortical potential evoked by visual stimuli several components can be distinguished. A very good example of VEP description was given by Cigânek (1961)/ Fig 4.1).

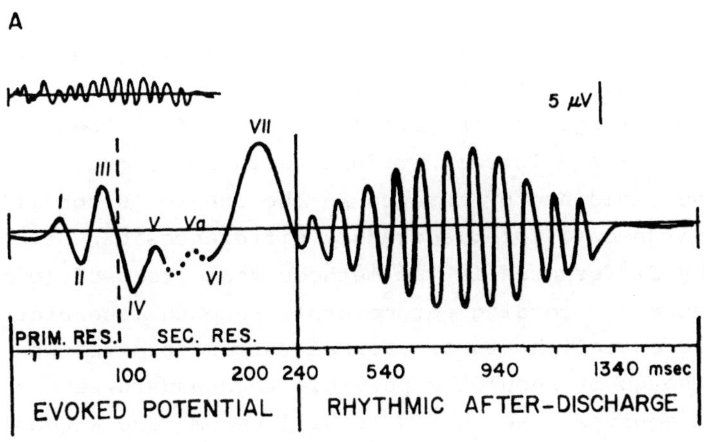

Fig. 4.1 Survey of the various phases and peaks of a visual evoked potential (Cigânek 1961; reproduced with permission)

He described a primary complex (0- ±100 ms), a more or less local and specific response, probably mainly from presynaptic activity of afferent thalamocortical neurones and originating in the IV layer of the visual cortex. A secondary complex can be distinguished between ±100- ±200 ms.

This is a more aspecific response, more generalised over the cortex. Multisynaptic activity from intercortical neurones, synaptic and dendritic, is supposed to be the major component. Initially much attention was paid to the differentiation between primary and secondary complex. Later on less attention was given to this dichotomy. No doubt, the first two or three cortical peaks are more stable, localised and specific, as the later components are aspecific, variable and influenced by attention and distraction. An afterdischarge between ± 250 and ± 1250 ms, initially described in the literature, did not get much attention later on and is seldom mentioned nowadays.

In the early fifties the technical facilities to average evoked potentials improved and commercial averagers became available. In the late fifties and early sixties great efforts were made to describe the normal response and to investigate the influence of age and the effect of attention. There are important inter- and intraindividual variations, that can partially be explained by changes in the subject's condition and the environment. Interlaboratory differences can often be explained by differences in the methods that are used (different stimulus and recording parameters). So each laboratory still has to establish its own normal values for its own method, although it should be possible to standardize the methods and equipment so that data will become exchangeable. There are age and sex differences, so that a group of normal reference test subjects should be selected carefully, matched for both sexes and the full life range. It is a common error to include only young men and women in the normal test subjects group. It is very important for evoked potential work to test in as standardized a manner as possible. The method of testing should be described in detail. Also environmental factors should be under control as musch as possible.
Finally, in the late sixties and early seventies, evoked potentials entered the clinical laboratories and proved to be an important diagnostic tool for several diseases of sense organs as well as the peripheral and the central nervous system.

Another major field of application is psychological testing and research.

2. ANATOMY AND PHYSIOLOGY

2.1 Anatomy of the eye

Embryologically the eye is part of the brain, i.e. a functionalized piece of ectodermic tissue. The exact shape of the eye is maintained by the combination of an intra-ocular pressure of 15 mmHg and the rigid structure of the eye wall (sclera and cornea). The light from the outside world is transmitted through the eye media (cornea, aqueous humour, lens and vitreous humour) towards the nervous part (Fig. 4.2).

The nervous part of the eye backwall, i.e. the retina, is a thin layer which is fed by two separate blood supply systems: the (inner) retinal vasculature and the subretinal choroid. The retinal layer has a kind of specialization which is to be seen most clearly in the existing fovea. This fovea is in the vicinity of the optical axis of the dioptric system formed by the eye media.

The optical system of the eye can be compared to an automatic TV camera: the pupil (compare the diaphragm) is automatically adapted to the average light level over a 20-fold range. The dioptric power is automatically adjusted from 10 cm to infinity (in early life) by changing the accommodation of the eyelens. It may be remarked that the overall dioptric power of the human eye equals approximately 60 diopters, 43 diopters of which are from the cornea and 18 (+10) diopters from the eyelens.

The aberrations of the dioptric system and the functional properties of the retinal regions are closely matched (by evolution), i.e. a high receptor density and small receptive fields of the retinal ganglion cells in the region of the optical center (fovea, macula) and a gradual coarsening of these features towards the periphery.

2.2 Physiology of the retina

Visual excitation, finally resulting in the visual evoked potentials, starts in the receptor layer of the retina (Fig.3.3). The receptors are the photosensitive cells, of which two types can be recognized, the rods for night-vision and the cones for day-vision.

The cones determine visual acuity and colour vision; furthermore they are mainly responsible for the sensitivity of the central part of the visual field and for the amplitudo of the pattern evoked potentials. Cone density increases enormously towards the fixation point. This implies that resolution, determining pattern recognition and visual acuity, increases towards the centre of the visual field.

As a consequence, pattern responses will be mediated more through the centre, as individual check size and contrast between the checks are smaller.

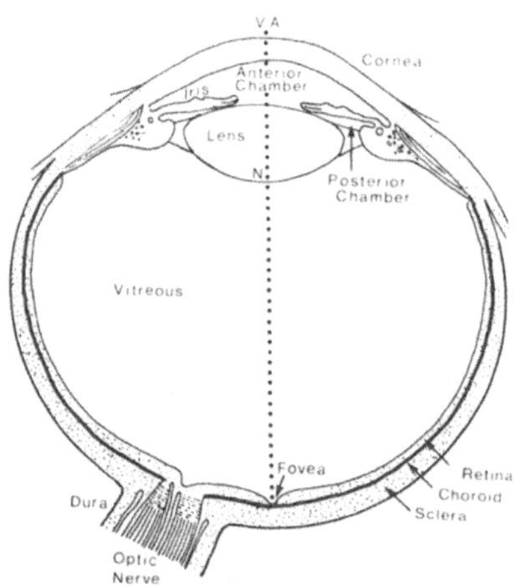

Fig. 4.2 Horizontal section of the eye.
VA = visual axis.
(Wolff, Anatomy of the eye and orbit).

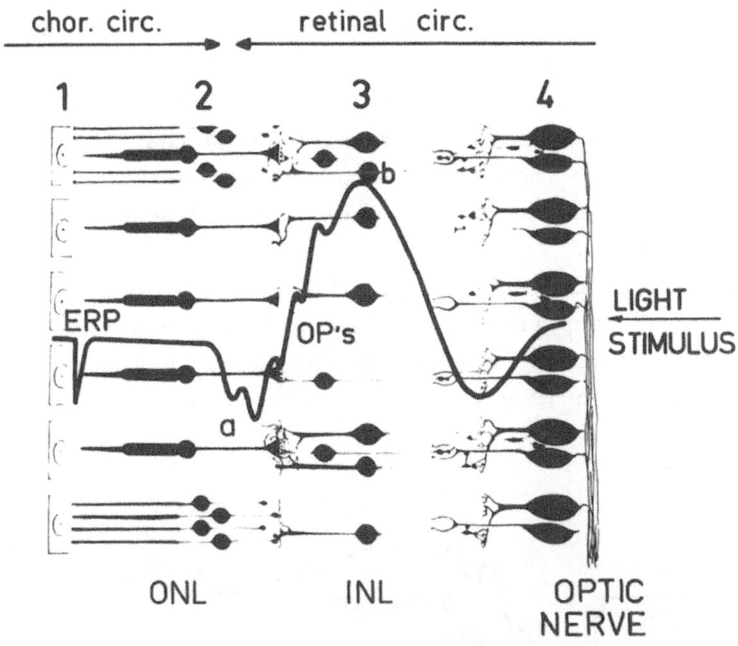

Fig. 4.3 1= pigment epithelium; 2= receptors; 3= bipolar cells; 4= ganglion cells; ONL= outer nuclear layer; ERP= early receptor potential; a = a-wave; b = b-wave; OPs= oscillatory potentials; INL= inner nuclear layer

In the outer segments of the photoreceptors, light energy is transferred into electrical energy, measurable as early receptor potential. Its clinical meaning is not proven yet.
In the inner segments, where the cell nucleus is located, the late receptor potential (i.e. the classical a-wave) originates. Together with the b-wave it forms the main part of the electroretinogram (ERG, Fig. 4.4).

The exact origin of the b-wave is not known. Very probably, it derives from the Müller cells. These are long glial cells, running from the outer nuclear layer of the retina to the surface of the proximal side. Clinically, it can be said that the ERG b-wave reflects the activity of the bipolar cell layer, the second order neurons in the visual process.
For clinical purposes it is important to know that the receptors are dependent on the choroidal blood supply, whereas the bipolar cell layer and the ganglion cell layer depend on the retinal circulation.
In deep retinal disturbances, both the a- and b-wave are lowered, whereas in purely superficial retinal disturbances like in recent occlusions of the retinal circulation, only the b-wave is lowered (van Lith 1980a)

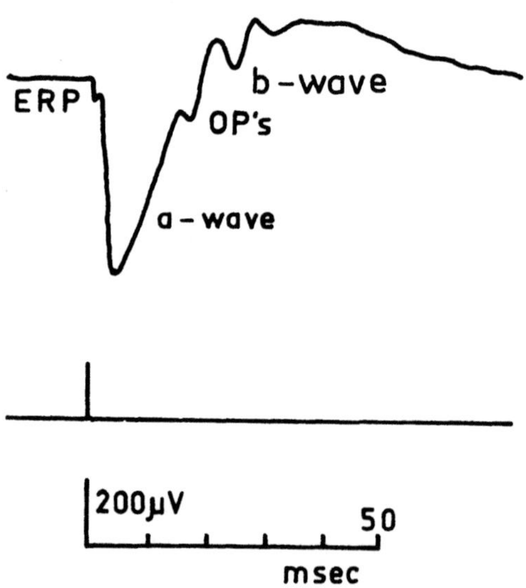

Fig. 4.4 ERP = early receptor potential
OPs = oscillatory potentials

The retinal ganglion cells are the third order neurons. Their axons constitute the optic nerve, which conducts the action potentials through the lateral geniculate body to the primary visual cortex. The activity of the ganglion cells is caused by the output of the bipolar cells and shaped by the amacryne cells. These latter cells are located in the proximal part of the inner nuclear layer (INL) and form a lateral interaction network. Its activity is mort probably reflected in the oscillatory potentials (OPs) which are superimposed on the ascending limb of the ERG b-wave. The OPs are affected in diseases of the superficial layers of the retina, among others in early stages of certain retinal circulatory disturbances.
ERGs evoked by pattern stimulation have been described recently by Persson and Wanger (1982), these responses too, are supposed to derive from this layer.

2.3 Physiology of nervous pathways

The nerve action potentials evoked in the retina are conducted by the optic nerve to the optic chiasma (Fig.4.5).

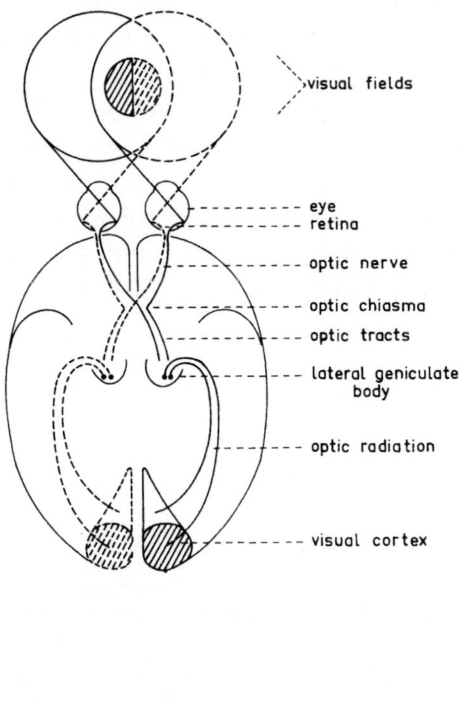

Fig.4.5

Here the fibres coming from the nasal halves of the retina of both eyes cross the median line. Fibres originating in the temporal halves of the retina of both eyes do not cross.
So retrochiasmatic information coming from the right and left side of the visual field is separated. Information coming from the right side of the visual field is joined and conducted by the left optic tract to the lateral geniculate body on the right side of the upper brain stem - thalamus complex.
The fibres originating in the cells of the retina synaps in the lateral geniculate body. From this relay station the optic radiation conducts the action potentials to the viscual cortex, primarily to the area 17. The neighbouring areas 18 and 19 are mainly visual associative areas. It is certain, however, that these areas partly participate in the VEP build-up. So impulses from the right half side of the visual field are projected to the occipital cortex of the left hemisphere.
On the other hand information coming from the left half side of the visual field terminate in the right occipital cortex area. The cortical representation of the central part of the retina, the fovea, is much longer, about 25x, as the peripheral part of the retina is. The upper gyrus of the calcarine fissure corresponds with the bottom side of the visual field; the lower gyrus of the calcarine fissure with the topfields.
It should be realized that major parts of the visual cortex are hidden in the calcarine fissure and are situated on the medial part of the occipital hemisphere and are not in direct contact with the skull. Only a minor part of the visual cortex is situated on the outer lateral surface of the brain and directly exposed to the skull . So it is difficult to explain what part of a scalp-recorded visual evoked potential correlates with what part of the visual cortex.

3. METHOD

3.1 Introduction

An important review paper concerning the method was written by Donchin et al. (1977). It is based upon proposals of an international working group chaired by the first author. Instructions for parameter description and method of testing are given.

As has been said before, each laboratory should establish its own normal reference values for its proper method. It is wise to become experienced with a restricted number of different stimulus and registration parameters. That means selecting a few of the many methods that have been - and will be - described in the literature.

3.2 Technical description of stimulation parameters.

3.2.1 Temporal characteristics of light modulation

The stimulation of the visual system can be continuous or intermittent.

Generally continuous stimulation is referred to as adaptation light. A more general definition of adaptation light would be the mean (effective) luminance level of a light source, irrespective of the temporal characteristics of the modulation. The term adaptation indicates that the sensitivity of the eye changes when the continuous luminance is changed. It is, therefore, important to specify its level adequately. Intermittent stimulation may be transient stimulation, or periodical stimulation, often called steady state stimulation. (Fig. 4.6).

Transient stimulation is defined as a stimulus with a duration which is very short with respect to its repetition rate, and to the time constants of the visual system. In practice the term transient stimulation applies to flashes and to stepwise changes of light. When the rate of repetition of the stimulus, either flash or any other change in luminance conditions, becomes higher than 6 to 10 Hz, the term steady state stimulation is used (cf. Sokol 1976)

This term indicates that the evoked responses fuse to a simple kind of (near) sine wave pattern.

3.2.2. Spatial characteristics of light stimulation

The term diffuse stimulation is used if no contrasts are present in the stimulus field. This can be achieved by using an adequate diffuser (transparancy, or reflecting surface). This kind of stimulation should be used in case of opaque eye media.

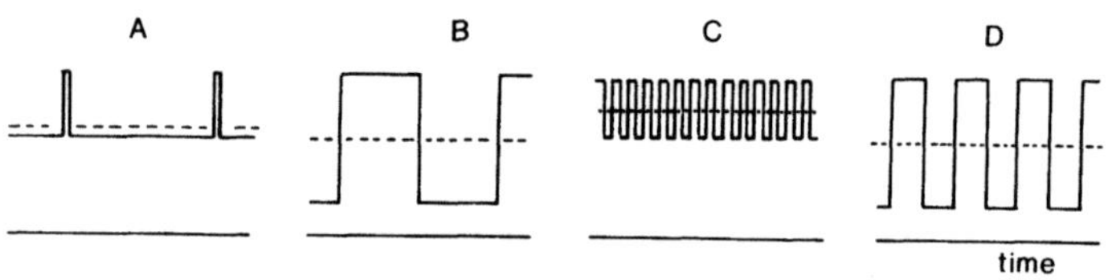

Fig.4.6 Temporal aspects of stimulation
a. transient pulse stimulation
b. transient wave stimulation
c. and d. steady state pulse and square wave stimulation respectively.

The alternative is to use a patterned field, which nowadays generally consists of a checkerboard pattern. One has in that case to specify not only the size of the checks (e.g. 15') but the contrast as well (e.g. 30%). The stimulus (Fig.4.7) may be the alternation of a diffuse field with a checkerboard and v.v. which is called pattern appearance-disappearance. Another possibility is to alternate the dark and the light checks, which is termed pattern reversal. In both cases the average (adaptation) level of the light intensity remains constant throughout the complete stimulation period.

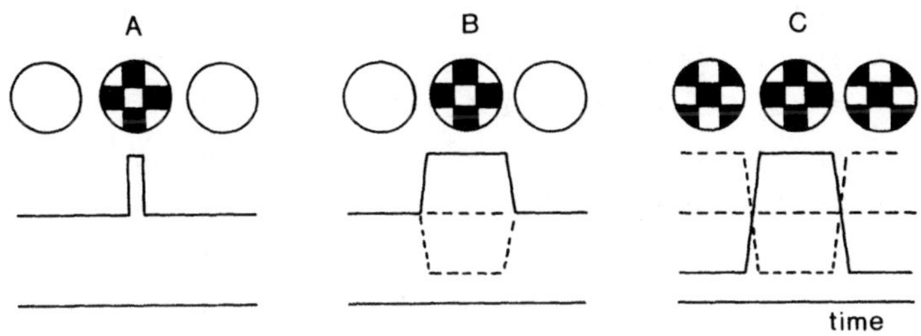

Fig. 4.7 a. flashed pattern stimulation
b. pattern appearance-disappearance (note that average light intensity remains constant)
c. pattern reversal.

The check size producing maximum responses with these conditions is of the order of 10'-20', which responses are sheer contrast responses. The third possibility is to put a diffuser with a checkerboard upon the flash unit and thus elicit a flashed pattern (or patterned flash), the advantages being the simplicity to create this opportunity from a normal flash unit and the unique and highly reproducable EP waveform (cf. Harter and White 1968; Sokol 1976). The criticism often given to this kind of stimulus is the mixture of luminance response and contrast response, but to a lesser extent this criticism also applies to the recently promoted use of pattern reversal stimulation with large check size of approximately 1° (cf. Regan and Richard 1971; Halliday et al. 1972; Spekreyse et al. 1973; Sokol 1976; Van Lith et al. 1978).
The size of the stimulus field has to be specified as well. It is known that the projection of the retina on the occipital cortex (cf. Brindley 1970) favours the central part of the

retina in the evoked responses. The minimum size of the field for obtaining a full response is, furthermore, dependent on the presence of a pattern. So, for diffuse stimulation under photopic conditions the central 12° of the retina mainly determines the evoked potentials and it may be considered the minimum field size (Van Lith and Henkes 1970; Kojima and Zrenner 1980); for patterned stimulation the central 3° are represented mainly (Spekreyse 1966; Rietveld et al. 1967) in the evoked response (decreasing with decreasing check size). It should be mentioned that for the convenience of the patient and for ensuring reliable fixation the field should at least be twice these sizes, because otherwise the fixation of gaze would have to be very stable. The stimulus field sould be kept more or less concentrically to the fovea, since it has been shown that the upper and lower half fields of patterned stimuli produce considerably different responses (Jeffreys and Axford 1972a, 1972b). Vertical half fields may be used to exclusively stimulate one cortical hemisphere. Special care has to be taken that the fovea itself is not stimulated in that case, because it has a double projection.

3.2.3. Luminance and contrast

In order to be sure that the cone photoreceptors are activated predominantly, an adaptation field (Ganzfeld, or large fields) with a luminance of at least 10 cd/m^2 (3 ft L) is needed. The stimulation is superimposed on this background. This can be realized by mixing background and stimulation with a semi-transmitting mirror, by projection of a stimulus on a background diffuser, or by any device employed for pattern stimulation (oscillating mirror device or TV monitor). The increase of the contrast will cause an increase of the magnitude of the response components up to a particular level, i.e. the saturation level (Fig. 4.8).

Further increase of the contrast may even produce a decrease of the response (presumably by adaptation effects), although the accompanying decrease of the peak latencies may still proceed.

The contrast should, therefore, not be fixed at a single level, but stimulus contrast conditions should span a factor of about 10. It will then be possible to estimate the slope of the response function as well as the saturation value of the response amplitude and, if possible, of the peak latencies.
It will be obvious that the pattern stimulus should be sharply depicted at the retina in order to obtain the optimum stimulation of the contrast response mechanism of the visual system. In effect at the optimum check size ($\sim$ 30') a small refractive error of the optical eye media yields a sharp decrease of the response magnitude. So one may conclude that an adequate ophthalmological examination should precede the evoked potential measurement.
Larger check size ($> 1^{o}$) reduces this effect, but with a subsequent mixing of luminance and contrast response effects in the evoked potentials (e.g. Harter and White 1968; Regan and Richard 1971; Bartl et al. 1978).
Another factor to be considered in the adaptation luminance level is the size of the pupil of the eye. If no mydriasis is applied, one should be aware that even at the photopic level the effective pupil area may decrease by a factor of two. As compared to a full size pupil under mydriasis the ratio is approximately a factor of eight (cf. LeGrand 1968). The effect will be noticed on both the amplitude and the latency of the various peaks of the VEP (Chelva and Van Lith 1982).

3.2.4. Frequency

The time interval between the stimuli is influencing the amplitude level considerably, even if the interval is longer than 1 second (cf. Jackson and Barber 1980). This applies to the secondary part of the VEP. If, on the other hand, the early oscillatory potentials are to be measured, a repetition rate of 6 Hz is often used, which is still adequate (cf. Cracco and Cracco 1978; Harding and Rubinstein 1981; Siegfried and Lukas 1981).

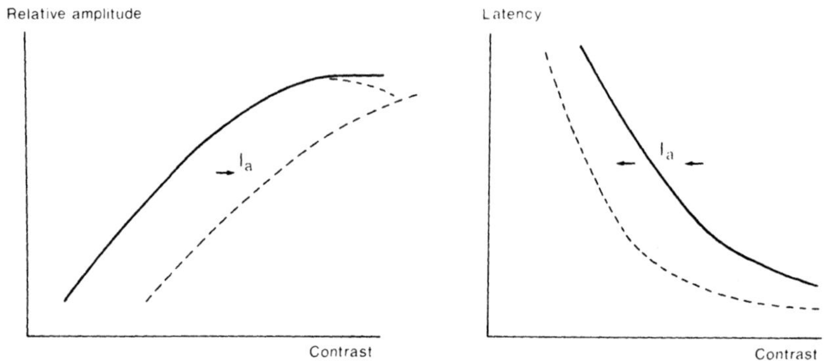

Fig. 4.8 Left part: Increase of contrast causes increase of VEP response amplitude up to a saturation value. Over-saturation may occur at still higher contrasts. Increase of adaptation level tends to shift the response curve to the right, i.e. a lower response at the same contrast.
Right part: Increase of contrast causes decrease of VEP latency. Higher adaptation level causes shift of the curve to the left, i.e. towards still shorter latencies.

3.2.5. Optimal stimulus parameters

A first effort to obtain some standarization was made by Thijssen and collegues (Thijssen et al. 1976; Braakhuis en Thijssen 1978), but their proposal also incorporated the conditions for the electroretinogram and the electrooculogram.

The conditions for optimal stimulation are dependent on the a priori choice of some parameters, i.e. the temporal and spatial characteristics of the stimulus pattern. The marginal condictions for obtaining VEPs and early oscillatory potentials are listed in Tables 3.1 and 3.2 respectively.

Table 3.1: Marginal stimulus conditions for VEP measurement.

background	color	white
	size	$>60^{o}$, Ganzfeld
	luminance	≥ 10 cd.m^{-2} (≈ 3ft L)
stimulus	repetition rate	≤ 1Hz
	color	white
diffuse (flash)	duration	≤ 10 ms
	size	$\geq 20^{o}$
	luminant energy	≥ 5 cd.m^{-2}.s
pattern (flash, pattern reversal, appearance-disappearance)	contrast	$\geq 30\%$
	check size	$\geq 20'$
	size	$\geq 10^{o}$

Table 3.2: Marginal conditions for early oscillatory potentials

background	See Table 4.1	
stimulus	repetition rate	≥ 6 Hz
	color	white
diffuse (flash)	duration	≤ 1 ms
	size	$\geq 10^{o}$
	luminant energy	≥ 100 cd.m^{-2}.s

3.3. Clinical aspects of stimulation

In Fig.4.9 several types of stimuli are reviewed.

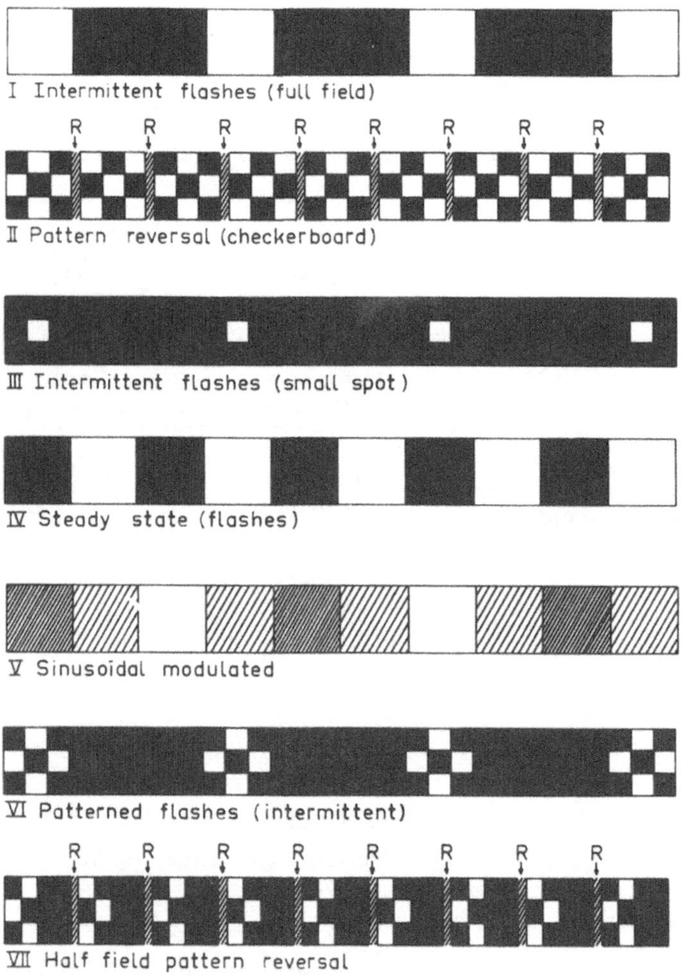

Fig. 4.9 Highly schematic review of several types of stimulation to evoke VEPs

I. Intermittent flashes (Fig.4.9I).
In the early days of evoked potential studies short duration, bright and diffuse stimuli, usually flashes by an electronic device, were used. In such a case large field stimulation - central as well as peripheral visual field - is applied. As a rule white light is used. Usually the first components are fairly well detectable, but the intra- and interindividual variability is rather high.
Stimulus parameters modify the response. Generally, increasing flash intensity decreases peak latencies and attenuates the amplitude of the peaks.

II. Pattern (checkerboard pattern reversal) (Fig.4.9II).
Later on, in the early seventies, patterned stimulation came to be used frequently. A pattern, usually a checkerboard pattern, is presented. Sudden reversal of pattern proved to be an appropriate stimulus to evoke a cortical potential. Peak latencies are shorter than they are in flash VEPs. This certainly applies to the secondary complex. However, it is sometimes said that a comparison cannot be made, because the wave forms evoked in a single person are clearly different for the two types of stimulation.
The advantage of this method is that the total amount of light does not change (it equals the adaptation light), and that the intra- and interindividual variability is less and the response wave form is stable. A disadvantage is that the first components are often not as well detectable as in flash VEP.
Currently these two are the stimulation methods most often applied. However, many modifications have been described; also stripes are used by some authors.

III. Small bright spot flash (Fig. 4.9III).
It can be necessary to stimulate the fovea selectively. In that case a "small bright spot" is used instead of "full field" stimulation (Hennerici et al. 1977).
Instead of full field (at least 30°), small field (about 1°) flashes are applied.

Only the central visual field (fovea) is supposed to be stimulated but straying of light in the optical media of the eye cannot be excluded. Moreover, it is difficult to assess the direction of the gaze of the patient during the experiment, which makes the method less suitable, if at all, for clinical routine examination.
Furthermore the response is mainly evoked by the central visual field; as this renders the response incomplete for this way of stimulation, the method cannot be regarded as very adequate.

IV. Steady state VEPs (Fig. 4.9IV).
Usually flashes as well as pattern reversals are delivered at low frequency (e.g. 1 Hz); in this way subsequent VEPs do not overlap. This stimulation method is known as "transient state" VEP.
At higher stimulus frequencies overlap of the responses occurs and no isolated evoked potential can be distinguished. However, spectral analysis of the EEG shows that e.g. 10 Hz stimulation evokes peaks in the power of the spectrum at 10, 13-25 and 40-60 Hz. So maximal steady state responses may be expected at these repetition rates (Spekreyse 1966).
The clinical application is not yet clear. Probably there are close similarities with the driving response of stroboscopic stimulation in routine EEG.
The advantage of applying stimuli at rates of >6 Hz is that the response wave form is very simple, so amplitude and latency are clearly defined and easy to measure.

V. Sinusoidal modulated light (Fig. 4.9V).
This is a more or less gradual variation of stimulus intensity. The light intensity is modulated at a certain frequency. Gradually changing light intensities are considered to be more physiological than flashes are. As has been said above (IV) no single VEPs can be detected. Correlation or power spectral analysis is needed to objectify the response. In the last decade this method got rather few applications.

VI. Patterned flash (Fig. 4.9VI).
A flash is delivered through a transparancy with a grid or checkerboard pattern. The latencies are shorter than in case of pattern reversal and the peaks are claimed to be easier to detect.
Generally this way of stimulation is not recommended, since a change in light intensity and a change in pattern occur at the same time, so that there is an interaction of two effects. In some laboratories a longer lasting flash, 100 ms ("semi-continuous") is used and a combined "on - and - off" effect is evoked.It is not clear yet whether there are advantages. In addition, there is another unknown effect since the "on" (appearance) and "off" (disappearance) responses will partially overlap.

VII. Half field stimulation (Fig. 4.9 VII)
Only one half, right or left, of the visual field is stimulated. One half of the checkerboard can be blinded, the other half is reversed. A half field of at least 15° is recommended. It is a surprising, but often confirmed fact that a large potential in the midline and ipsilaterally was evoked by half field pattern reversal stimulation. Stimulation of the right half of the visual field of the right eye evoked a potential mainly over the right occipital area (Halliday and Michael 1970). However, recent publications show that this asymmetry is undoubtly true for mean values in normal populations, but certainly not in all individuals. By half field stimulation many normal subjects show no significant asymmetry; others show the previously described asymmetry in favour of the ipsi-lateral hemisphere, but also in a number of subjects the contrary is found. Whether the response will be found homo- or ipsilaterally depends on stimulus parameters (Halliday and Michael 1970). When a pattern flash is applied the largest response is found over the contralateral hemisphere (Jeffreys and Axford 1972a, 1972b). Even both asymmetries can be found in the same response (Van Lith et al. 1981). The conclusion is that the findings highly depend on stimulus conditions.

They also determined how often asymmetries can be found in normal individuals with half field stimulation or in hemianopic patients with full field stimulation.
Half field stimulation is recommended to detect hemianopsias. Undoubtly a number of hemianopsias are not recognized if not examined by half field stimulation (Rowe 1982).

3.4. Recommended choice of stimulus parameters

There are good reasons - as has been argued on before- to become familair with only a limited number of stimulation methods.
It should be remembered that for all methods of stimulation in a laboratory, normal values have to be established. This holds also for different light intensities and check sizes, if used!
For ophthalmological purposes, checkerboard patterns in a reversal mode are mostly applied. The responses obtained are rather simple, since, in general, only P100 is measured.
This is, of course, an enormous limitation, the more so when only one size of checkerboard pattern is used. The advantage is, however, that normalcy can much better be defined and interpretation is more easy.
Flash and pattern reversal stimulation, both large field, will be discussed more in detail.

A. Flash

- Intensity: flash stimuli are often delivered by a gas discharge tube with an electrical energy of about 1 Joule. Flash intensity can be specified in Candela or Lumen, as explained in the Appendix. The tube should be in front of the eyes at a distance of 25-30 cm.
  The stimulus intensity effect will of course be influenced by the background light intensity. It is neither necessary nor pratical to test the subject in absolute darkness, but the background light intensity should be standardized. No light from outside may shine into the room, because great differences between sunny days and rainy weather will occur.

With low background illumination levels the retinal scotopic system is mainly stimulated. This causes the IV and V peak (Cigánek nomination) of the VEP to decrease and the VI to increase. On the other hand in light-adapted situations (mainly photopic stimulation) a well-developed peak V appears.

- Duration: with the commercially available flash equipment short duration flashes can be delivered usually. The flash duration should be ≤ 10 ms.
- Colour: as a rule white light is used. For special indications in ophthalmology one might use coloured light (i.e. red or blue) which influences the response, so normal reference values should again be available (see Appendix).
- Field size: usually diffuse large field ($>30^{\circ}$) stimulation is performed.
- Frequency: regarding a response duration of 500 ms or even longer a repetition rate of 1 Hz should be considered to be the maximum so as to prevent overlap of responses and recovery phase problems. As a result of regular stimulation changes in the responsiveness of the central nervous system occur, which has been described as "habituation", "cyclic conditioning" and so on. So from a theoretical point of view stimulation "at random" (at irregular intervals) should be recommended. Nevertheless in pratice a regular 1 Hz stimulation frequency is often used.In short time stimulation as used in VEP averaging the theoretical objections are supposed to be of minor importance.

B. Pattern reversal.

- Type: "checkerboard patterns" are most commonly used.
- Intensity: overall luminance should be specified for the total field, e.g. 30 cd/m^2. Moreover the black-white contrast should be mentioned, preferably in percentage of the average light level. This can be expressed by the formulae (cf. also Appendix)

$$\frac{\text{difference in luminance}}{\text{total luminance}} \times 100\%$$

Sometimes the dark to light ratio is specified, i.e. the luminance ratio between the dark checks and the light checks (e.g. 1:14).

- Reversal speed: there are two reversal methods. A mechanical one: a mirror driven by a galvanometer; with this system a reversal time of <10 ms can be performed. This should be regarded as the most appropriate reversal speed.
  On the other hand a changing TV monitor picture is used. Depending on the number of sweeps that are necessary to build up the picture, it takes 20-40 ms before the whole field from top to bottom has been changed. In our opinion this reversal time is too long for adequate measurement of peak latencies.
- Field size: a field size of about 30° and a check size of about 1° are recommended.
  The check size can be varied. The potential evoked by small checks is somewhat more complex than in case of a larger check size. Changing the check size can be performed to test objectively the visual acuity. However, this is a much more complicated and time-consuming way of visual acuity testing than the conventional way by letter reading cards is. Of course in all situations where no visual acuity testing is intended, refraction anomalies should be corrected before pattern reversal VEP testing is performed. As a rule, for this purpose the subject's own glasses will be used.
- Frequency: the same frequency conditions as for flash VEP are recommended for pattern reversal VEP.

C. General remarks.

- Choice of the eyes: keeping the eyes open has many advantages, but keeping the eyes closed during flash stimulation is more convenient for the patient. However, it should be realized that the light levels become unpredictable, as does the direction of the gaze. In all sorts of patterned stimulation the eyes should be open. In a number of neurological patients, and also in testing response/attention relationships in psychology, binocular stimulation will be sufficient.

However, with many other neurological and with all ophthalmological indications it is necessary to test both eyes separately.

- Steady state of alterness: it is wise and easy to avoid periods of sleep. Since recording of two VEPs (e.g. 2 x 60 trials to average) by a stimulus frequency of 1 Hz will last about 2 min, subsequently, recording of both eyes in this way does not exceed 5 min. Thereafter a rest and recreation period should be included before doing more tests.
- Contamination: one should be careful with the stimulation (c.q. reversal) equipment. To prevent contamination of the VEP with an AEP, no sound, especially no clicks, may be heard simultaneously with the visual stimuli.
- Contamination by the electroretinogram (ERG) artefact is discussed under the heading Normal Short Latency VEP.
- It is recommended not to deliver stimuli in periods of possible artefacts, e.g. as caused by unrest in children and in anxious and mentally disturbed patients. This certainly holds true if no automatic artefact rejection system forms part of the averaging system.

3.5. Recording

A. Electrodes

The same electrodes as in routine EEG can be used. It is strongly recommended to use the standardized international electrode positions known from the 10-20 system. Unfortunately in a number of papers aberrant electrode positions are used, in which case the localization of the electrodes has to be specified in detail.

B. Montages.

There is no international or national standardized lead program. In the majority of the papers 1 to 4, sometimes even more, leads are used simultaneously with a minimum of only one midline position (Oz). In our opinion in neurology 4 positions (O1,O2,Oz and Pz) are to be considered a good number, and at least O1 and O2 should be included.

Concerning the reference electrode many opinions exist. The references most often used are: Fz or "linked ears" (A1/A2). Also used are: single ipsi- or controlateral ear, mastoids, common or average reference and even bipolar leads (Fig·4.10).

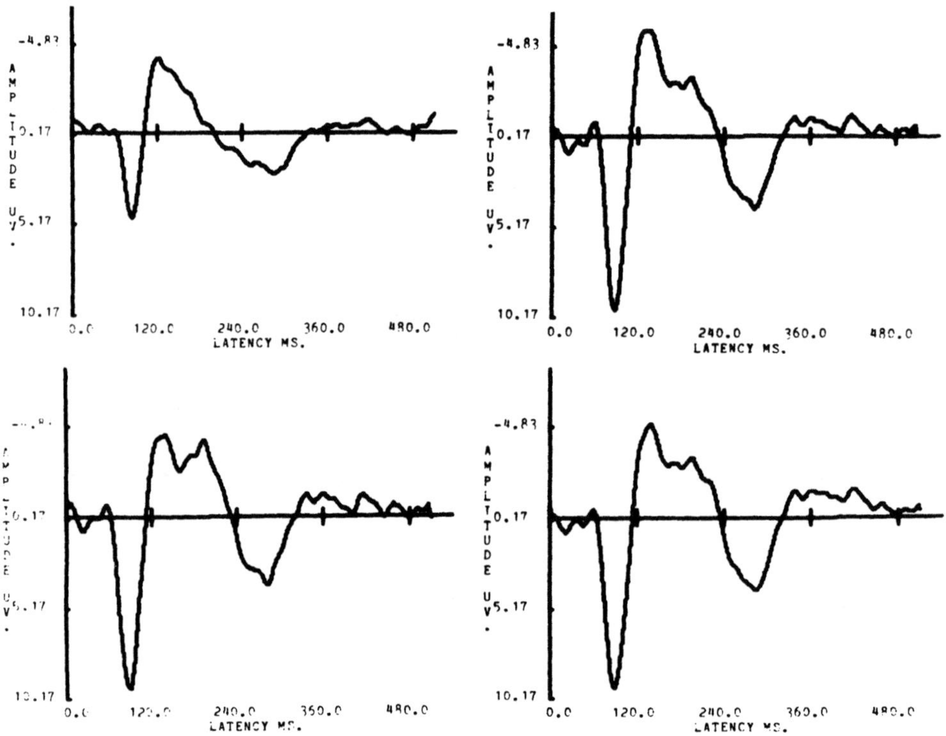

Fig. 4.10 Four leads pattern reversal evoked potential
a. Pz - Fz; b. Oz - Fz; c. O1 - Fz; d. O2 - Fz
amplitude: -4.83 to +10.17 uV
time base: 0.0 to 500.0 ms

In ophthalmological application electrodes over the hemispheres are necessary at least when asymmetric visual field defects are present; otherwise, only medial leads may suffice.
With these leads, as a rule, no important contamination with muscle activity occurs. Of course eye artefacts, especially if appearing synchronously with the stimuli, can be troublesome.

Simultaneous recording of eye movements may be useful to detect the risk of eye movement contamination. Sometimes eye artefact rejection or elimination systems are used.

B. Registration.

In EEG registration positivity downward and negativity upward is worldwide practice, since the early days of electroencephalography. Unfortunately this is unlike the common practice in physiology, electrotechnology and ophthalmology. Neurologists prefer and recommend to join the clinical routine EEG practice of positivity downward - negativity upward. It should be noted though, that in a number of papers, not only from physiological laboratories, the reverse is done.
To prevent misunderstanding and errors in the figures positivity and negativity should be clearly indicated (Fig. 4.11).

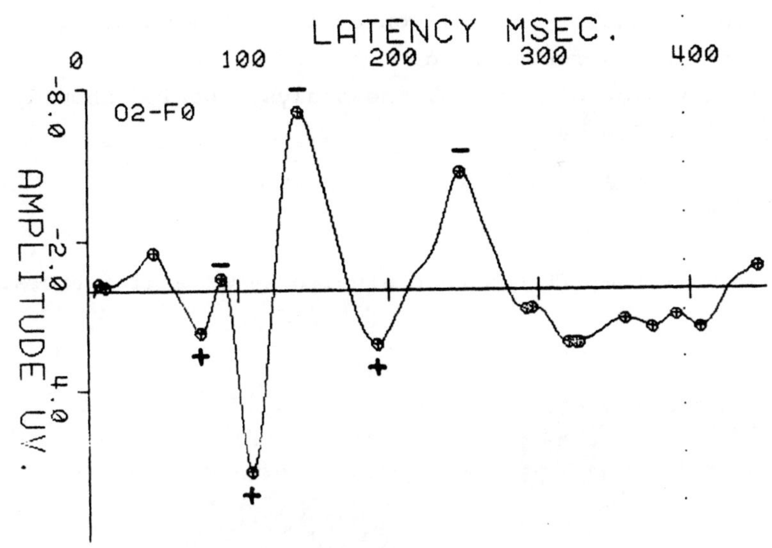

Fig. 4.11 Visual evoked potential (flash evoked)
amplitude : -8.0 to + 8.0 µV
time base : 0 - 400 ms
Top directed peaks are negative (O2 negative to Fz) and bottom directed peaks are positive (O2 positive to Fz)

4. METHOD OF ANALYSIS

Only a few remarks are given about the technical aspects of recording and averaging. More detailed information is presented above.

The bandwidth of the recording system should be 0.5-70 Hz (-3dB). Recording with an upper limit of 30 Hz (-3dB) certainly effects the rather fast earliest cortical peaks. If, in case of pattern reversal stimulation, attention is paid to the later components only (e.g. P100), a 30 Hz (-3dB) filter may be acceptable.

Of course calibration is necessary when calculating the amplitude of the response.

Averaging of 60 or 120 stimuli responses will usually be sufficient to record a well-shaped cortical visual potential from the scalp. An analog-digital conversion (ADC) sampling frequency of 500 Hz is recommended.

Regarding the duration of the VEP, the analysis period should at least be 400 to 500 ms.

5. LOCALIZATION

The topography of the VEP has been studied by several authors.

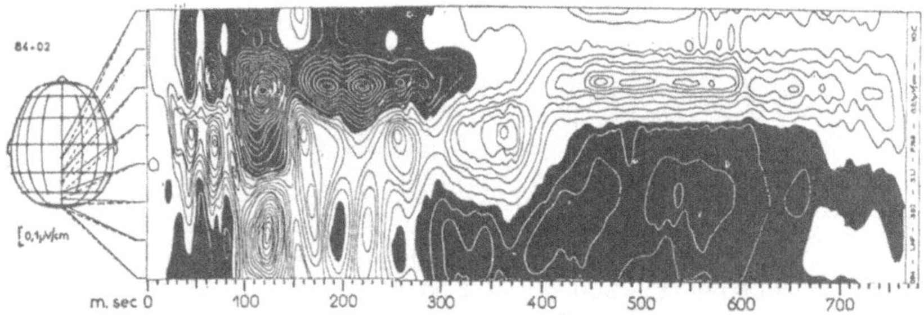

Fig.4.12A Spatio-temporal map of the VEP to flash (from Rémond and Lesèvre, 1965).

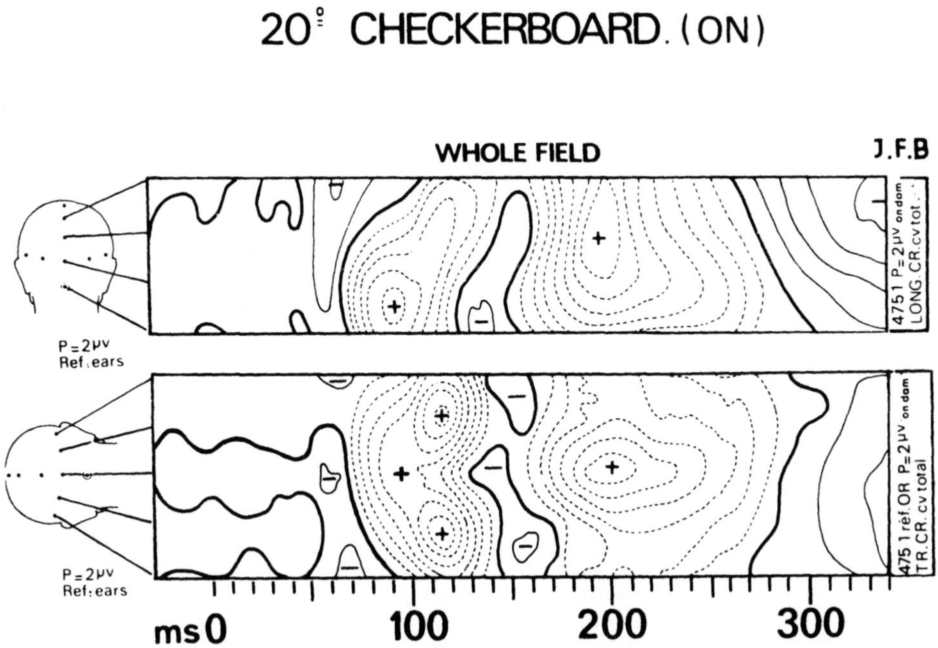

Fig. 4.12B Spatio-temporal map of the VEP to pattern reversal (Lesèvre, 1976)

Generally the highest amplitude of the VEP is found in the midline at the inion or at Oz. However, there are important differences. Certainly not all peaks have their maximum at that area. Generally the early cortical components are more localized with a maximum in the occipital area (over the visual cortex); others, especially the peaks of the secondary complex, are more widespread, even with a maximum at the vertex. An unexpected finding was the ipsilateral VEP attenuation in case of homonymeous hemianopsias (Vaughan et al. 1963; Blumhardt et al. 1977).

More research has to be done to the localization of the cortical dipoles that contribute to the build-up of the VEP (see Jeffreys and Axford 1972a, 1972b; Van Lith et al. 1981).

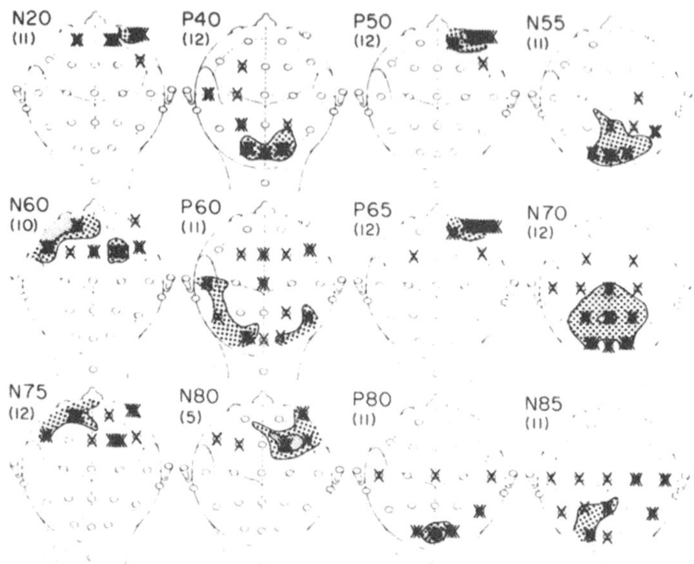

Fig. 4.13 Topography of VEP components. Isopotential maps were constructed (from Allison et al. 1977)

Topographical analysis can also be applied to study the difference between macular and peripheral vision (Lesèvre 1976).

6. DESCRIPTION OF THE NORMAL VEP

6.1. Peak latencies and amplitudes

Unfortunately the peak nomenclature is awkward and unstandardized. Initially, especially for the flash VEP, the nomenclature as proposed by Cigánek (1961) and Gastaut and Régis (1965) was used. At least six subsequent peaks should be distinguished: I (pos), II (neg), III (pos), IV (neg), V (pos) and VI (neg). It should be noted that in normal subjects not all peaks can always be identified.

Nowadays it is common practice to denominate the peaks by a letter-number combination. N stand for negativity, P for positivity. The number gives the rounded off normal latency value in ms (Fig. 4.14).

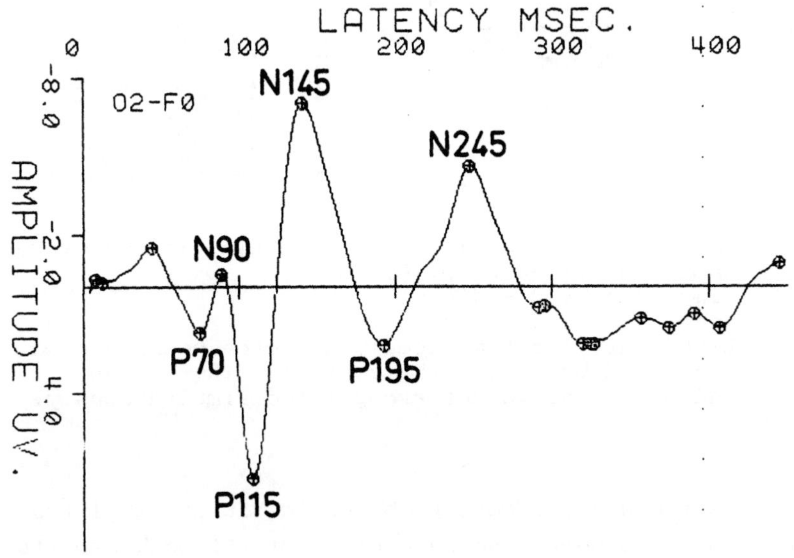

Fig. 4.14 Visual evoked potential (flash evoked)
amplitude : -8.0 to + 8.0 µV
time base : 0 - 400 ms
The peaks are enumerated according to their polarity and latency.

For instance P100 stands for a positive peak with a mean normal latency of 100 ms.

Of course the figure depends on the normal reference values that a particular laboratory uses with testing.

Latencies are measured from the beginning of the stimulus to the maximum of the peak (Fig.4.15) .

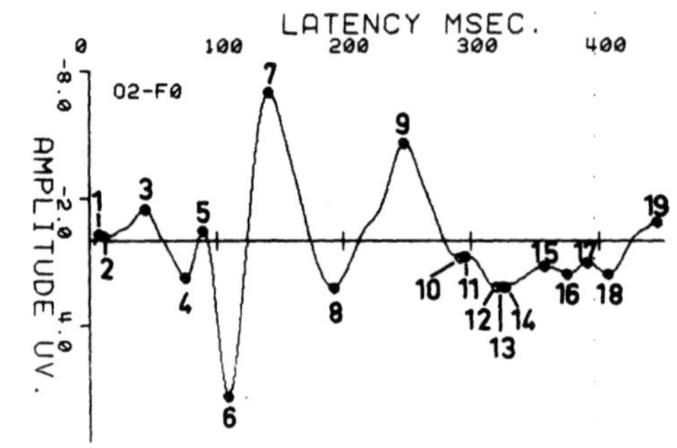

02-FF

TOPNR.	LATENCY	AMPLITUDE
1	9.5 MSEC.	-0.31 UV.
2	13.5 MSEC.	-0.22 UV.
3	45.2 MSEC.	-1.55 UV.
4	77.0 MSEC.	1.74 UV.
5	90.9 MSEC.	-0.50 UV.
6	110.7 MSEC.	7.27 UV.
7	142.4 MSEC.	-7.05 UV.
8	194.0 MSEC.	2.21 UV.
9	249.6 MSEC.	-4.64 UV.
10	293.2 MSEC.	0.75 UV.
11	297.2 MSEC.	0.72 UV.
12	321.0 MSEC.	2.16 UV.
13	327.0 MSEC.	2.11 UV.
14	328.9 MSEC.	2.11 UV.
15	350.7 MSEC.	1.16 UV.
16	376.6 MSEC.	1.55 UV.
17	392.4 MSEC.	0.99 UV.
18	408.3 MSEC.	1.52 UV.
19	446.0 MSEC.	-0.90 UV.

Fig. 4.15 Left part : visual evoked potential (flash evoked)
amplitude : -0.8 to +0.8 µV
time base : 0 - 400 ms
lead : O2 - Fz
Right part : automatically detected peaks are indicated by a dot; the corresponding latency and amplitude values are printed simultaneously.

Amplitude measurements are more problematic. In EEG amplitudes are measured peak-to-peak. The same goes for VEP peak amplitudes (Fig. 4.15). But should a P 100 amplitude be measured in relation to the preceding or the following negative peak? And what to do when the preceding (or following) negative peak cannot be identified?

Another approach could be to measure absolute amplitude values in reference to the zero line.

However, the question is how to specify the zero line?

There are three ways to answer this question:

a. the technical zero level as found in calibration;
b. the mean value of the total VEP;
c. the mean value of 100 ms averaged signal preceding the stimulus.

All approaches have their pros and cons. We feel that the "mean value of the total VEP" is a level of almost zero; it is also most easy to determine. It is preferable, therefore, to measure peak amplitudes in relation to this zero level.

6.2 Reproducibility

It is becoming common practice to test the reproducibility of the response by performing signal averaging twice (2x 60 signals averaged) and to register both VEPs in the same figure. In case of good reproducibility both VEPs should overlap (Fig 4.16).

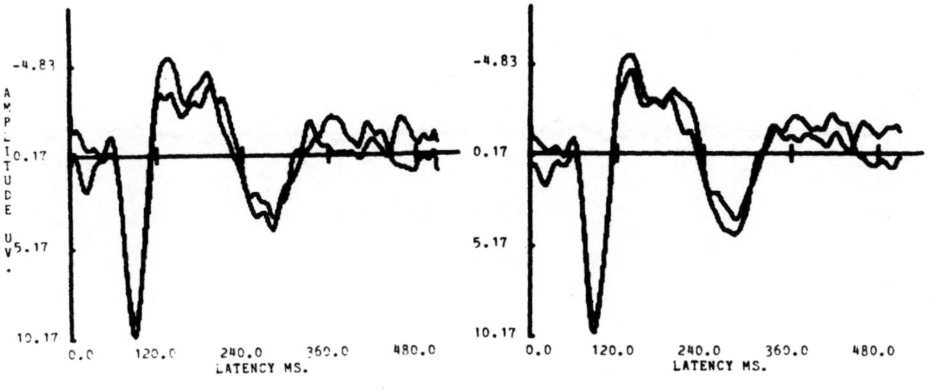

Fig. 4.16 Visual evoked potential (pattern reversal evoked) left part: O1 - Fz; right part: O2 - Fz Two subsequent averaged signals do overlap.

If no good overlap exists, the validity of the VEP may be doubted. In such a case an additional security check can be done, i.e. recording a "zero"-response. In the same way as done before 60 signals of 500 ms duration are averaged but without giving any stimulus. Theoretically such a "zero"-response should be a "response" at about zero level (Fig. 4.17).

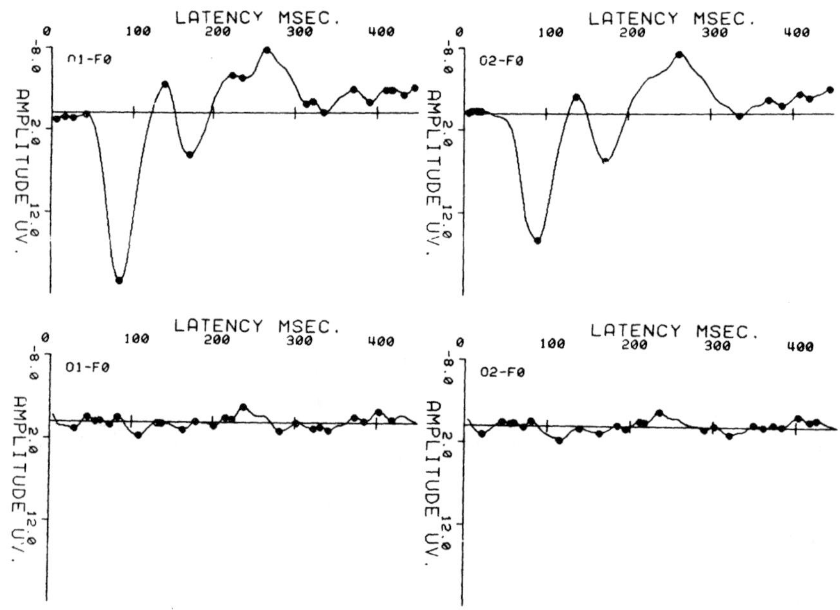

Fig. 4.17 Top : Visual evoked potential (pattern reversal evoked)
left part: O1 - Fz; right part: O2 - Fz
Bottom : result of averaging signals without any visual stimulus.
No response occurs, no activity is recorded.

When performing this check up it is not always easy to cancel all "background" EEG activity by averaging.
Especially in case of a large amplitude very regular alpha rhythm residual "background" EEG activity can still mask or distort the averaged VEP (Fig.4.18).

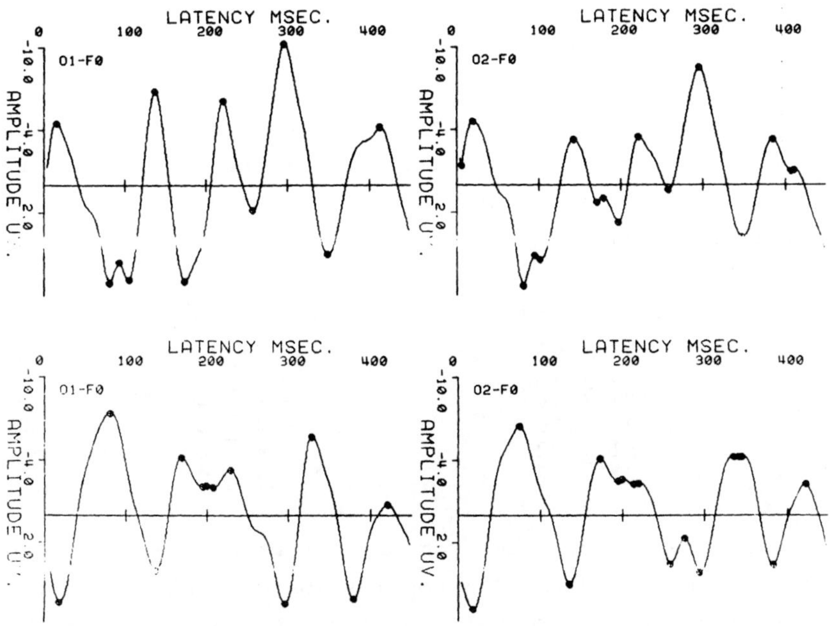

'ig. 4.18 Top : VEP trial (pattern reversal evoked) left part: O1 - Fz; right part: O2-Fz The response (if any) is not very characteristic.

Bottom : result of averaging signals without any stimulus shows that EEG background activity is not satisfactory eliminated.

7. NORMATIVE VALUES

7.1. Peak polarity, latency and amplitude

The VEP has already been described in the late fifties / early sixties (Cobb and Dawson 1960; Ellingson 1960). As has been said before a good example of VEP description was given by Cigánek (1961) (Fig. 4.1). He distinguished six main peaks, alternating positive and negative ones and he denominated them I to VI.

Since then many papers have been published presenting normal values, which, however, were sometimes more or less contradictory. Cigánek (1961) found the following normal values (means and 1x standard deviations):

	latency in ms	amplitude in µV
I	50 ± 13	+ 2 ± 2
II	70 ± 14	- 1 ± 3
III	103 ± 14	+ 9 ± 4
IV	146 ± 20	- 4 ± 3
V	191 ± 22	+ 6 ± 4
VI	233 ± 32	- 5 ± 3

A well-documented study came from Bennett et al. (1971). They examined 50 normal adults. A monopolar montage (with an ear reference) and a flash frequency of 1 Hz was used and 128 responses were averaged. They found the following normal values:

polarity	latency in ms	amplitude in µV
+	46 ± 12	1 ± 2
-	69 ± 15	1 ± 2
+	108 ± 9	7 ± 3
-	147 ± 13	5 ± 3
+	196 ± 20	9 ± 4
-	246 ± 29	6 ± 3

In our laboratory (SLV) the following method is used:

-flash	intensity	400 Lux (at a distance of 25 cm before the closed eyes
	duration	10 ms
	colour	white
	field size	"full" ($>30^{\circ}$)
	frequency	1 Hz
-checkerboard	pattern reversal	
	intensity	20 $cd.m^{-2}$ (total field, 45 cm in front of the open eyes)
	contrast	black:white = 1:14 (87%)
	field size	$33^{\circ}20'$
	check size	$1^{\circ}20'$
	reversal time	7 ms
	frequency	1 Hz
-montage	Pz-Fz, Oz-Fz, O1-Fz,	O2-Fz
-bandwidth	0.5-70 Hz (-3dB)	
-averaging	sample frequency	500 Hz
	number of sweeps	64 or 128
	averaging duration	450 ms

Posthuma and Visser (1982) found the following normal reference values:

	latency in ms		amplitude in µV	
P	CB	F	CB	F
+	54 ± 6	72 ± 16	2 ± 2	3 ± 3
-	66 ± 6	92 ± 17	2 ± 2	2 ± 3
+	96 ± 4	113 ± 31	9 ± 3	4 ± 3
-	129 ± 7	144 ± 44	3 ± 3	1 ± 2
+	166 ± 13	196 ± 23	3 ± 1	4 ± 3
-	219 ± 28	244 ± 33	2 ± 2	2 ± 2

P = polarity; CB = checkerboard pattern reversal evoked;
F = flash evoked.

(see also Fig. 4.19).

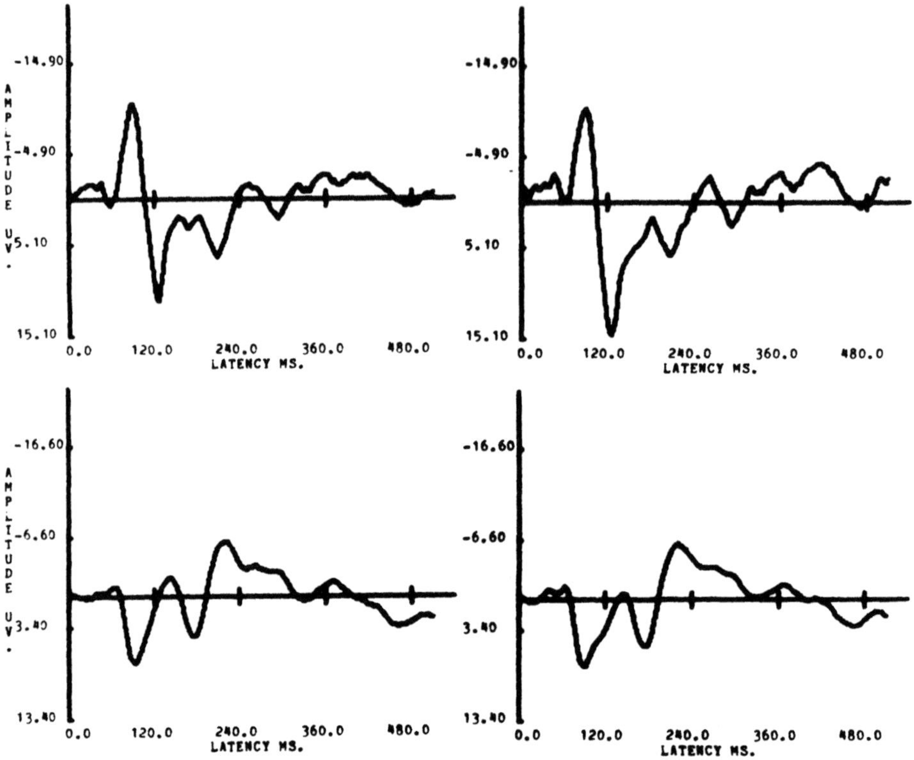

Fig. 4.19 Visual evoked potential; top: flash evoked; bottom: pattern reversal evoked.
Left leads: O1-Fz; right leads: O2-Fz

7.2. Peak recognition

Not all peaks can always be detected in all normal adults. The first two peaks, which are part of the primary complex, are often not easy to recognize in case of pattern reversal stimulation. The P100, the major positive peak with a latency of about 100 ms, is the one that is most easily recognizable and most stable.

In 20 normal adults Posthuma and Visser (1980) detected the following number of peaks:

polarity	checkerboard pattern rev.	flash
+	2 (10%)	15 (75%)
-	6 (30%)	18 (90%)
+	20 (100%)	18 (90%)
-	19 (95%)	16 (80%)
+	13 (65%)	13 (65%)
-	10 (50%)	12 (60%)

It proved that in checkerboard pattern reversal stimulation the early peaks could be recognized only in 10-30% of the subjects; in flash VEP this was 75-90%. On the other hand the major positive peak at about 100 ms could be recognized in all (100%) pattern reversal VEPs and in 90% of the flash VEPs. The later components were recognizable in 50-75% of the subjects, irrespective the stimulation method.
Sometimes a certain peak can be easily detected in one hemisphere and cannot be well recognized in the other. It may be difficult to determine whether this is pathological or not.

7.3. Short latency VEP

It is often supposed that P50 is the first cortical component. Compared with the first cortical component of the SSEP at 20 ms this is rather slow. A few authors mentioned short latency (brain stem?) VEPs. Gastaut et al. (1967) described a P25-N40 complex, followed by a P60 and N80, the latter being the peak I and II of the primary complex.
Cracco and Cracco (1978) found early oscillations: a N14, P18, N21, P24 complex. Allison et al. (1977) only found a N20-P40 complex and Goff et al. (1979) described a P40 only.
Recently Harding and Rubinstein (1981) could recognize a P20, N25, P35 complex in 28 out of 30 normal volunteers. The maximum of the response was found at the T4 and T8 scalp positions.
The short latency VEP evoked by flashes may be contaminated by the response of the eye (electroretinogram, ERG). The ERG consists of a cornea negative-positive wave form with latencies.

of 15 ms and 35 ms respectively, with photopic conditions (normally used in VEP experiments), followed by a slow positive wave, which is, however, often disturbed by the blink reflex. The amplitude of the photopic ERG (cone ERG) is of the order of 100 µV. It may be remarked that the frequency range of the short latency VEP is from 70 to 140 Hz (Cracco and Cracco 1978), so adequate high pass filtering of the VEP will remove possible interference of the ERG ($\lesssim$40 Hz). What may remain is the contamination of the VEP by the so-called oscillatory potentials of the ERG (100-150 Hz), which, however, are much smaller in amplitude than the main ERG wave form components (< 20 µV).

7.4. Symmetry

Concerning latencies there are no systematic asymmetries. Several authors state that normal latency asymmetries should not exceed 5 to 10 ms (for P100). For late components of the VEP greater normal asymmetries can probably exist.
However, systematic amplitude asymmetries are also described, even in case of full field stimulation. The left/right amplitude asymmetry is described and said to be similar to the alpha rhythm asymmetry. According to several authors amplitude asymmetries up to 25-40% are normal.
For our method, previously described (Posthuma and Visser 1982), we found the following normal left/right asymmetries:

	latency in ms		amplitude in µV	
	CB	F	CB	F
+	-	3 ± 1	-	3 ± 1
-	3 ± 2	3 ± 1	<1 ± <1	3 ± 1
+	2 ± 3	2 ± 1	1 ± 1	2 ± 1
-	5 ± 4	4 ± 5	1 ± 1	4 ± 5
+	3 ± 3	4 ± 6	1 ± 1	4 ± 6
-	3 ± 3	4 ± 3	1 ± 1	4 ± 3

Normal reference values for left/right hemisphere differences.

	latency in ms		amplitude in µV	
	CB	F	CB	F
+	-	3 ± 3	-	1 ± 1
-	1 ± 1	4 ± 4	1 ± 1	1 ± 1
+	2 ± 2	5 ± 5	2 ± 1	1 ± 1
-	7 ± 10	3 ± 3	2 ± 1	1 ± <1
+	6 ± 3	7 ± 5	1 ± 1	1 ± 1
-	6 ± 6	12 ± 4	1 ± 1	1 ± 1

Normal reference values for left/right eye differences.

From these data we conclude that P100 (flash) has a normal latency asymmetry of 5 ms (interhemispheric) and 15 ms (interocular). For P100 (pattern reversal) 7.5 ms hemisphere latency asymmetry and 6 ms ocular asymmetry was found.

7.5. Sex

The influence of sex in the EEG has often been neglected. However, lately several authors have proved for sex to influence the VEP. Shearer and Dustman (1980) published the following data (in ms):

	P50	N65	P100	N150	P200
female	46	66	95	153	201
male	52	68	95	157	209

On the basis of his data Halliday (1981) stated that VEP latencies in women are 3.5 to 3.9 ms shorter than latencies in men. Indeed the VEP amplitudes in men are 4.6 to 4.9 µV larger than in women. However, concerning amplitudes the opposite has been described as well (women having larger amplitudes than men).

7.6. Age

Early publications on the influence of age on VEP are those by Ellingson (1960) and Gastaut and Régis (1965). Other extensive and well-documented studies were done by Dustman and Beck (1969), Laget et al. (1976), Blom et al. (1980) and Barnet et al. (1980).

In neonates the response is rather flat and simple: only one large negative wave or biphasic response (N1 156 ± 26 ms and P1 180 ± 20 ms) (see Fig.4.20).

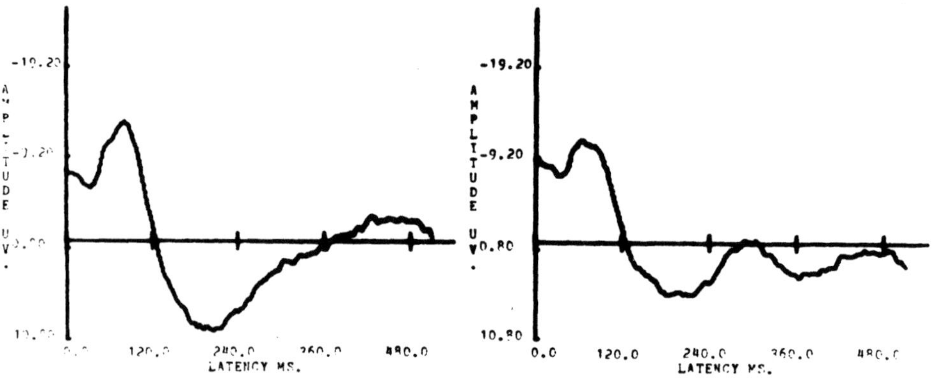

Fig. 4.20 Visual evoked potential (flash evoked): simple neonatal biphasic response.
Left: O1-Fz, right: O2-Fz

There is a great variability. Laget et al. (1977) described in neonates only one large positive wave with a latency of about 190 ms. After two months of age more differentiation is seen: a large negative wave appears at 175 ms. It should be stressed again that there is a great variability. We have seen nearly adult-shaped VEPs in babies only a few months old. On the other hand, as Mellor and Fielder (1980) state, absence of VEP in the first months of life does not necessarily predict abnormality. These authors recommend to repeat the test at the age of 6 months before coming to a final conclusion.
In the course of the first year of life the differentiation proceeds. More peaks can be recognized and the variability decreases. It was found that the distribution over the cortex increases with age, while the latencies of the peaks decrease gradually.
Generally speaking it can be said that at the age of 5 to 6 (other authors say even at the age of 3 to 4) the shape,

latency and amplitude values reach about adult values.
The amplitude increases from 0-6 years of age. However, the course of the amplitude increase is more variable. It has been found that the amplitude increases from 1 month up to 5-6 years of age. Thereafter a slight amplitude decrease was observed, again followed by an increase after the age of 13 to 14. After the age of 16 the amplitude remains about the same.
Relatively few papers are published on amplitude with respect to old age. Celesia and Daly (1977) examined a group of normal adults from 18 to 79 years of age. The amplitude did not change significantly, but the latencies increased with aging. Generally it can be said that until the age of 55 to 60 the influence of age can be neglected. The latency increase over 50 years of age is found to be greater in women than in men.
Kjaer (1980a) published upper limit normal reference values for men and women of several age groups (P100):

men	10-19 y	115 ms	women	20-59 y	107 ms
	20-49 y	110 ms		60-69 y	110 ms
	50-69 y	120 ms			

The influence of age is more significant for the first three peaks (primary complex) than for the late components. In case of pattern reversal stimulation the influence of age on the VEP is greater with a small (12') than with a larger (48') check size (Sokol et al. 1981).

7.7. Attention

Changes in the VEP caused by changes in alterness level were described as early as the sixties. The VEP amplitude increases by high attention levels and decreases by distraction. This effect is much greater for the late components than it is in the primary complex. The changes in the VEP run more or less parallel with the level of alterness effects in the EEG.
Two kinds of stimuli can be distinguished. These are more profoundly investigated in case of AEP. When presenting frequently a 500 Hz sound and rarely a 600 Hz sound after instructing

the subject to count the number of rare stimuli, the 600 Hz AEP (in this example) will have a larger amplitude than the AEP to the common 500 Hz sound. This phenomenon has been described as "task relevancy" effect.
Similar effects can be expected to occur in the VEP.
In case of repeated examination habituation occurs. This habituation is a common effect in multisynaptic responses of the central nervous system. Habituation has a greater effect on the amplitude than on the latency or shape of the response.
By conditioning experiments the response can be facilitated or inhibited. It is not always easy to predict whether potentiation or habituation will occur.
Amplitude changes by habituation or conditioning affect mainly the early components of the VEP. On the other hand habituation has been denied and has been described as a pupil constriction effect only.
In literature experimental methods and environmental situations vary widely. So it can be concluded that facilitation and inhibition might occur. One should always be aware of these effects.

7.8. Sleep

There is no important effect of light sleep (stage I) on the VEP. Some amplitude decrease can occur. In deeper slow sleep stages (II-III) the amplitude of the VEP increases. Largest amplitudes are found in sleep stages with K complexes. During REM sleep ("active sleep") all peaks show small amplitudes. Although the effect of sleep seems to be rather small, it is nevertheless important to avoid possible sleep effects in VEP testing. The subject should be kept awake during the experiments. Certainly in all cases where no sleep effects in particular are to be investigated VEP testing should be performed with the subject awake and in state of alertness.

8. CLINICAL APPLICATION

8.1. Ophthalmology

8.1.1. Introduction

In ophthalmology measurement of the VEPs is only of limited value. Most abnormalities can be diagnosed much better and much more easily by other means, such as measurement of visual acuity and visual field, by splitlamp examination or fundoscopy. For the differential diagnosis of retinal diseases or defects, the VEP is even less suitable; this can better be done by means of retinal electrical responses in electrooculogram (EOG) and electroretinogram (ERG) or by fluorescein angiography. Therefore, this group of diseases will not be dealt with here.

The VEPs are most important in case of an unexplained visual loss (lowered visual acuity or visual field defect) or in case the interpretation of a disc anomaly provides difficulties.
In either case measurement of the response to light flashes is too rough a method whereas application of pattern stimulation reveals much better results. Apart from being a more refined stimulus, the wave form of these responses is more simple and less variable. This implies that detection of certain peaks and measurement of their amplitudes or latencies are more reliable (see Fig.4.21).
In principal, only four main types of pattern reversal VEP can be discerned:
- normal amplitude and latency (Fig.4.21,A):
- normal amplitude and increased latency (Fig.4.21,B);
- lowered amplitude and normal latency (Fig.4.21,C):
- lowered amplitude and increased latency (Fig. 4.21,D):

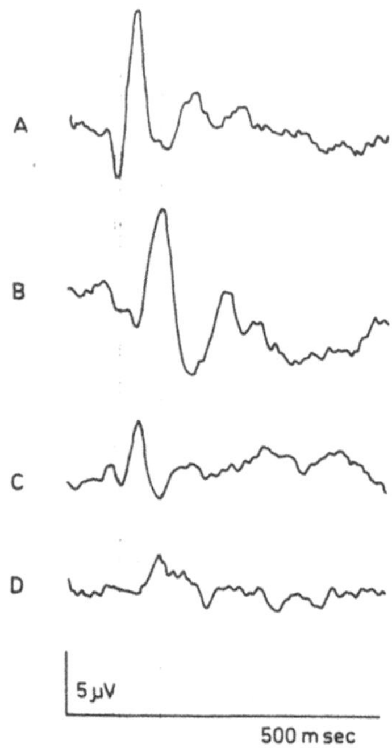

Fig.4.21 Four main types of pattern reversal VEP (see text).

An absent or non-detectable response is, in fact, a response of which the amplitude is so much lowered that the latency cannot be measured. Sometimes, also wave form alterations are found, but these have not yet been determined that they can be properly used.

With only four features of abnormalcy, diagnostic possibilities are rather limited, but this inconvenience can be compensated by combining them with the four fundoscopic manifestations of the optic disc which can generally be encountered under pathological conditions.

These are:

- a normal disc
- disc edema
- the excavated disc
- atrophy of the disc

The number of possible variations now becomes (4 times 4 makes) 16. A further improvement can even be obtained by comparing the results of the pattern responses with the psychophysical results (visual acuity and visual field), since in some diseases pattern responses are more disturbed than can be detected by psychophysical methods (e.g. in multiple sclerosis), whereas in others the reverse is true (often in optic atrophies)

It is often asked whether the diagnostic power of pattern stimulation can be increased by application of various check sizes and contrasts between the checks. This is very likely to be true. A problem is, however, that the smaller the check sizes are and the lower the contrast is, variability increases and responses are more likely to be disturbed by little abnormalities in the optic media or retina. Moreover, normal reference values are not so well established. When abnormal responses are obtained with small checks and low contrasts, disturbances of optic media or retina can also explain the abnormal values and, therefore, should be carefully excluded. With larger checks and higher contrasts, however, this is less critical. In such latter circumstances abnormal responses point more to organic disturbances of the visual pathways, provided that the central retinal area does not show clearly visible gross abnormalities.
The various disanomalies will now be dealt with, providing some additional data to facilitate differential diagnosis. It may be stressed, however, that a careful ophthalmological examination should not be omitted in all pattern evoked response measurements.

8.1.2. Normal fundi

Patients, most often referred to VEP testing by opthalmologists, are those with visual disturbances and normal fundi or presumed normal fundi. The cause of these unexplained visual disturbances can be:

a. Acute retrobulbar neuritis:
  - acute onset, more often in young female adults, usually onesided.
  - central scotoma, spontaneous remission within some days.
  - VEPs relatively strongly disturbed (compared to psychophysics); always lowered and delayed, sometimes also in the normal fellow eye.

b. Demyelinating diseases (multiple sclerosis):
  - silent plaques may cause delayed responses of normal amplitude; but amplitudes being more dependent on the stage of atrophy. For the frequency of abnormal responses see 8.6.

c. Compression of optic nerve fibres in or before the chiasm:
  - onset at any age gradual, but sometimes experienced as acute by the patient; furthermore gradual decrease after the onset.
  - peripheral field defect, more often temporal.
  - no recovery or only slight temporary recovery.
  - especially processes before the chiasm may imitate an optic neuritis.
  - relatively strong disturbances of the VEPs, lowered and delayed.
  - sometimes also in the other eye (process near the chiasm).

When in case of an unexplained visual loss, one of the afore mentioned symptoms is not typical of an optic neuritis and when delayed responses are found, a compressive lesion is explicitely to be excluded (Van Lith et al. 1982).

d. Processes beyond the chiasm:
  - the VEPs can be asymmetric in amplitude over both hemispheres, in accordance with hemianopic field defects,

but they are relatively less disturbed compared to the former group.*

- no difference could be found between intracranial tumors and circulatory disturbances in this group by means of VEP (Crevits et al. 1983).

e. Leber's hereditary optic atrophy:
- acute onset, on either side frequently occurring in young male adults.
- central scotoma, no remission, transition to the atrophic stage.
- in acute stage resemblance to optic neuritis; VEPs lowered and delayed.

f. Intoxications, mainly tobacco-alcohol neuropathy, sometimes ethambutol:
- gradual onset on either side, usually male adults.
- centrocoecal scotoma, remission possible but usually not seen.
- delayed and lowered VEPs in the intoxication state, but also ERG abnormalities in advanced cases of alcohol neuropathy (Van Lith 1980a, Posthuma and Visser 1982).

g. Acute ischemic optic neuropathy:
- sectorial disc edema, later on optic atrophy (see 8.1.3.).

h. Chronic ischemic optic neuropathy (see optic atrophy).

i. Stationary congenital retinal dysfunctions (colour disturbances and nyctalopias):
- existing from birth, on either side.
- diagnosis via colour tests, dark adaptation curve and ERG.
- various abnormalities of the VEP, dependent on the type of dysfunction.

* In hemianopic field defects, also half field stimulation can trace abnormalities which by full field stimulation could be missed (Müller-Jensen et al. 1981; Camacho et al. 1981). But also half field stimulation is not always helpful (Maitland et al. 1982).

j. Amblyopia:
- one-sided, existing probably from the first year of life.
- normal VEPs with large check sizes; lowered and even delayed responses with small check sizes. This combination, however, is not typical of the amblyopia alone, but can also occur in refractive anomalies and medial cpacities. which, however, can be excluded by routine examination.

k. Psychogenetic visual loss, both by malingering or by neurosis:
- gradual or acute onset, usually on either side.
- normal VEPs, even with very small check sizes and low contrast.

It is remarkable that in most examples of this group of visual disturbances with normal fundi, diagnosis can be made with large check sizes; only when the responses are normal, should smaller check sizes also be applied in order to detect the abnormalities of an amblyopia or to establish the lack of abnormalities in the psychogenetic group.

8.1.3. Disc edemas

A second important group, where measurement of pattern VEPs may contribute a great deal to the differential diagnosis, is that of the disc edemas. We can distinguish:

a. Choked discs:
- not often seen by ophthalmologists with a visual loss as the only symptom.
- when combined with other signs of increased intracranial pressure, examination of the VEP not opportune.
- VEPs relatively little disturbed in comparison with those in demyelinating diseases or compressive lesions.
- with the conventional Halliday pattern reversal method, normal VEPs possible even when visual acuity is clearly disturbed. (Van Lith et al. 1982).

b. Papillitis:
- VEPs and clinical picture are the same as in acute retrobulbar neuritis.

c. Ischemic disc edema:
- acute onset, older age, one-sided.
- sectorial disc edema and visual field defect.
- in acute stage lowered and delayed VEP combined with a lowered EOG slight rise and a normal ERG (Brudet et al. 1983)
- within some months the transition to the atrophic stage with some improvement of the visual functions, a normal EOG and a sectorial disc atrophy.
- in this stage VEPs dependent on the extent of the defect.

d. Hypertensive disc edema (not often seen):
- VEPs probably lowered and delayed.
- EOG and ERG abnormalities according to the retinal circulatory disturbance (Van Lith 1980b, 1980c).

e. Drusen of the optic disc:
- onset gradually, young adults, on one or either side.
- no remission, gradual decrease.
- VEPs little or not disturbed, like in the choked disc, even when there is a visual loss (Brudet and Van Lith 1981).

f. Juxtapapillary chorioretinitis with collateral disc edema:
- acute onset, young adults, one-sided.
- sector-shaped defect in the visual field.
- VEPs relatively little disturbed.
- remission with a stationary field defect, VEPs dependent on this effect.

In the group of disc edemas it seems that the VEP is not so much disturbed by the edema itself, provided that the nerve fibres are not yet involved. This might be the case in increased intracranial pressure, drusen of the disc and in the juxtapapillary chorioretinitis.
In disc edema by a circulatory disturbance it is very likely that the axons are involved, since lateron always an atrophy of the disc is seen. In this case the VEP appears to be more disturbed, especially in the acute stage.

8.1.4. Excavated discs

Of not so much importance is the registration of the VEP for the excavated discs because the diagnosis of a glaucoma being principally made by ophthalmoscopy, examination of the fields and measurement of the intraocular pressure. But it is important to know that glaucoma, both the hypertensive and the normotensive type, may increase latencies (Huber and Wagner 1978).

8.1.5. Disc atrophies

As yet, a very difficult and extensive field for investigation is the group of disc atrophies. As a whole, they are in general characterized by lowered VEPs, the degree of which being dependent on the loss of sensitivity in the central part of the visual field. Though increased latencies may occur, they are not so expressive as seen in demyelinating or compressive lesions.

Ikeda et al. (1978) stated that amplitude reduction is an important indicator for axonal involvement. A problem of amplitude measurements is the large variability as compared to latency measurements. In one-sided disturbances the other healthy eye can be used as a reference, implying that the smaller variability of the interocular difference can be applied.

In long standing, district optic atrophies, when no occipital responses can be obtained, a differential diagnosis by means of VEP is, of course, completely impossible. In these cases the patient's history provides extremely useful data, while the retinal electrical responses may be of help, since they have to be normal in glaucoma, MS, compressive lesions and intracranial processes, whereas they may be abnormal in circulatory processes, tobacco-alcohol neuropathies and in some hereditary optic atrophies (Van Lith and Vijfvinkel-Bruinenga 1978; Van Lith and Henkes 1979).

If the disc atrophy has not progressed too far and clear responses are obtained, two rules of thumb, which have been mentioned before, can be applied.

Relatively strong disturbance of the VEP as compared to psychophysics and/or strong delays combined with rather good amplitudes point to a demyelinating process (on both sides) or a compressive lesion(usually one-sided) as the cause of the disc atrophy. No or only small delays, together with small amplitudes and/or relatively good VEPs in clearly atrophic discs and lowered visual acuity are more suggestive of an axonal degeneration in the optic nerve. This will be found in chronic or late stages of circulatory disturbances and intoxication neuropathies, in hereditary optic atrophies (e.g. Leber's atrophy), as well as in late or end stages of MS. These are the rules we can go by as yet.

8.1.6. Some general ophthalmological remarks

a. Neurophysiologists, applying pattern VEPs as a diagnostic aid, should not ignore abnormalities of the eye itself, which may also disturb pattern responses.
   - refractive anomalies (amplitude reduction) and medial opacities (both amplitude reduction and latency increase), that are more disturbing as the pattern elements are finer (Chelva and Van Lith 1982).
   - decreased retinal illumination (latency increase and amplitude reduction) by sunglasses, narrow pupils (glaucoma therapy!!) and yellow brown cataracts.
   - retinal diseases, that cause a lowered sensitivity in the central part of the visual field; atrophic processes have a greater influence on the amplitudes, edematous processes probably have a greater influence on the latencies.
   - abnormalities of the optic disc, especially glaucoma.
b. There are some diseases which may be easily overlooked or which may imitate neurological entities. Careful ophthalmoscopy while dilatring the pupils is necessary, sometimes also fluorescein angiography of measurement of the retinal electrical responses.

Some examples are:

- retinopathia centralis serosa:
  usually a one-sided edema of the central retinal area, with an acute onset in young adults, resembling optic neuritis.
- cone dystrophies:
  in this group foveal abnormalities may be very difficult to recognize, so that only the temporal pallor of the disc will be seen.
- normotensive glaucoma:
  by the normal intraocular pressure the proper diagnosis can be missed.
- tilted disc syndrome:
  a stationary connatal disc anomaly, usually on two sides, often combined with partial sectorial retinal atrophy in the lower nasal quadrants. The bilateral upper temporal field defects resemble a chiasm syndrome. Unfortunately, pattern responses may also be delayed in the tilted disc syndrome; in those cases only ophthalmoscopy will reveal the proper diagnosis (Wijngaarde en Van Lith 1981).

c. From this survey it may be clear that VEP is often decisive in establishing the cause of an unexplained visual loss; furthermore VEP may be of help in case the cause of a disc edema is not clear. In the assessment of the cause of optic atrophies registration is sometimes worthwhile, but may be of no avail in other cases. Interpreting VEP results, one has to be very careful not to overlook abnormalities of the eyes themselves, especially those abnormalities which also cause delayed responses, for an enoveous diagnosis of an MS may then be wrongly affirmed.
   Another misleading finding may be that of delayed response, when confirmation of a possible diagnosis MS is asked, whereas really a compressive lesion is present.

8.2. Local cerebral hemisphere lesions

Liberson and Scott (1964) and Larson et al. (1966) studied VEP abnormalities in patients with cerebrovascular accidents.Other early publications are those of Jonkman (1967) and Oosterhuis

et al. (1969) The VEP is often delayed or attenuated on the controlateral side.
In neurological patients with cerebral lesions, with or without hemianopsia, the homolateral VEP latency may be delayed and the amplitude may be decreased. However, with the aid of only VEP, precise localization of the lesion within the hemisphere is difficult to establish. More recently it was found that in case of cerebrovascular accidents amplitude asymmetries ($>$ 25%) are found more often than latency asymmetries (Thorwirth 1981).

In case of cerebral tumours several VEP abnormalities have been reported; amplitude attenuation, increased latencies, changes in polarity and form of the complex or even absence of VEPs (Schneider 1968; Crighel et al. 1976). VEP testing can probably be of value for the prognosis.
Abraham et al. (1975) examined a number of patients with acute blindness by occlusion of the basilar artery. In case of absence of VEP the prognosis proved to be worse compared with patients in whom the VEP was present.

In summary VEP testing is of little value for localizing and identifying ischemic and space occupying lesions. However, it is valuable in detecting late sequelae (this has been described in neuropediatrics) and in establishing prognosis.

8.3. Organic psychiatric syndromes

Harden et al. (1973) described 16 children suffering from late infantile neuronal lipidosis (Batten's disease) in which the VEP was often absent. However, when the VEP could be recorded, the amplitude was often extremely large and even visible without any averaging.
Patients with a Down syndrome (mongoloid idiocy) show increased latencies and also increased amplitudes (Gliddon et al. 1975).

Pampiglione (1977) presented an excellent review of VEP abnormalities in inborn errors of metabolism. Especially in

Tay Sachs disease VEP abnormalities are found. EEG, VEP and ERG abnormalities do not always run parallel. The VEP can be absent, but extremely large amplitudes may be observed.

Lee and Blair (1973) published data of 3 patients suffering from M. Jakob Creutzfeldt. From one of these patients VEP data are available. Already early in the course of the disease latency increases and amplitude decreases.
Rossini et al. (1979) also described one patient with M. Jakob Creutzfeldt with a delayed P100. Instead of a normal value between 90-120 ms a 160 ms latency was found. The response could not be evoked during paroxysmal activity in the EEG.
Visser et al. (1976) examined 19 patients with senile and presenile dementia (Alzheimer type). The latencies of the peaks II, III and IV (Cigánek nomenclature) were delayed and the amplitudes of the II and III peak were increased. Laurian et al. (1977) found either normal or absent VEPs in dementia.
In Huntington's disease the VEP latency is described to be normal, but the amplitude is found to be attenuated (Ellenberger et al.(1978). These findings, to be sure, are not always obvious (Oepen et al. 1982).
There are only few studies on VEP in chronic alcoholics. The results suggest a decrease in late component VEP amplitudes. Alcohol withdrawal is followed by recovery of these amplitudes. In chronic alcoholics interhemispheric asymmetries are decreased (Coger et al. 1976; Dustman et al. 1979; Porjesz and Begleiter 1979; Salamy et al. 1980). The same was found by Posthuma and Visser (1982). Especially in chronic alcoholism showing a Korsakow syndrome latencies can be increased. The abnormalities can possibly be reversible.
In summary in diffuse cerebral atrophy dementias the VEP is often found to be abnormal. Increased latencies and even absence of VEP are reported. The severity of these abnormalities probably correlates to the stage of the disease.

8.4. Non-organic psychiatric syndromes

There is a large field of literature regarding VEP applications in functional psychiatric disorders (Shagass and Schwartz 1965; Speck et al. 1966; Heninger and Speck 1966; Begleiter et al. 1967; Bostem et al. 1967; Floris et al. 1967; Shagass 1972). However, the results are rather diverse. The VEP can possibly be used to predict worsening or improvement of the disease. However, a clearcut clinical application has not yet been found. In summary up till now real sound clinical applications in the diagnosis of non-organic psychiatric syndromes have not been found.

8.5. Child neurology

In children with learning disorders the amplitude of the peak with a latency of about 200 ms is found to be delayed, especially in the left parietal region (Conners 1971; Preston et al. 1974). Also, Njiokiktjien et al. (1977) found delayed latencies in 58 children with learning disorders. The amplitude proved to be increased.

The same can be concluded from a paper by Symann-Louett et al. (1977), who found that children with learning disorders showed most peaks between 100 and 350 ms, whereas in normal children most peaks were found before 200 ms.

The VEP can be used to estimate the conceptional age of premature neonates (Graziani and Weitzman 1972). Ellingson (1960) already described longer latencies in premature than in "a term" neonates. 85% of the children suffering from asphyxia neonatorum showed delayed and irregular VEPs. There appeared to be a correlation between the degree of VEP abnormality and the severity of the clinical symptoms (Hrbek et al. 1977).

In case of "idiopathic respiratory distress syndrome" there are inconsistent VEP abnormalities (Hrbek et al. 1978). These abnormalities, caused by perinatal stress factors, can be traced for a long time afterwards. Marcus (1977) found that "high risk" children in the first three months of life showed longer VEP latencies than normal birth risk children.

Gambi et al. (1980) found in perinatal respiratory distress children even at the age of 9 months longer VEP latencies than in children without perinatal problems. The neurological condition at birth can be expressed by the Prechtl score (Prechtl 1977). A score ≤ 36 is an indication of a non-optimal neurological condition of the neonate. From papers of Poortinga et al. (1982) and Visser et al. (1981,1982) it appears that children with a non-optimal neonatal neurological score showed, even at the age of 5 years, longer VEP latencies than children with a better condition at birth. VEP proved a better indicator for late sequelae of birth trauma than EEG. The same has been described for SSEP by Laget et al. (1976).
In summary VEP testing can be used in neuropediatrics to objectify prematuritas and perinatal neurological problems. Late effects of a non-optimal neurological condition at birth can be detected throughout the first years of life.

8.6. Multiple sclerosis

8.6.1. Introduction

After Baker et al. (1968) found an abnormal SSEP in 68 out of 91 multiple sclerosis patients, Richey et al. (1971) described delayed VEP latencies.
These abnormalities are correlated with the severity and stage of the disease. The amplitude was found to be abnormal. Feinsod et al. (1973) reported that already in the very early stage of the disease the VEP can be abnormal, even in patients without any clinical sign of visual field defects. Subclinical VEP abnormalities can predict future visual defects (Walsh et al. 1982).

8.6.2. Choice of stimulation

Pattern reversal is a more effective method of stimulation than flash is (Halliday et al. 1972; Collins et al. 1978). Duwaer and Spekreyse (1978) found the flash VEP abnormal in 3 out of 18 patients with multiple sclerosis and the pattern VEP in 12 out of 18 patients. However, the opposite can be true as well;

patients have been reported with normal pattern reversal and abnormal flash VEPs (Neetens et al. 1979). Hennerici et al. (1977) and Oepen et al. (1981) described "small bright spot" flash, stimulating the fovea only, which should be more effective than full field flash stimulation or pattern reversal.

	Hennerici et al. 1977		Oepen et al. 1981	
	PR	SS	PR	SS
proven MS	81%	94%	75%	76%
probably MS	67%	94%	51%	80%
possible MS	43%	78%	31%	54%

PR = pattern reversal; SS = small spot

Diener and Scheibler (1980) compared in the same way multiple sclerosis patients with and without symptoms of retrobulbar neuritis:

	pattern reversal	small spot
MS with retrobulbar neuritis	100%	100%
MS without retrobulbar neuritis	57%	73%

Wilson and Keyser (1980) also published that full field flash is not the most effective stimulus. However, intermittent patterned flash and pattern reversal seem to be of equal value in MS diagnosis.
Aminoff and Ochs (1981) found "pattern onset" to be more effective than pattern reversal.
As a rule VEP tests are first choice in MS diagnosis. It can be completed by SSEP (with a preference for tibial posterior of peroneal nerve stimulation) and BAEP. The choice depends upon the expected localization of "plaques". In case of suspected MS with complaints of visus it is not only important to confirm the visual abnormalities; it is even more important to prove another process localization within the central nervous system (see Green and Walcoff 1982).
The detection of a clinically silent "plaque", which indicates multiple localizations, is necessary to confirm the diagnosis

of MS (Dorfman and Cummins 1979; Kjaer 1980c). In a group of 29 MS patients Kjaer (1980c) found the VEP to be abnormal 18, the SSEP 16, and the BAEP 15 times.

8.6.3. Incidence

There are many publications on the correlation between VEP abnormalities and the clinic. A few papers will be reviewed:

	Hennerici et al.1977	Matthews et al.1977	Shahrokhi et al 1978	Kayed et al.1978
proven MS	81%	75%		92%
probable MS	67%	58%		61%
possible MS	43%	38%		39%
all MS			57%	
MS + retrob. neuritis			91%	100%

	Kjaer 1980b	Lehmann and Soukos 1982
proven MS	100%	68%
probable MS		50%
possible MS	70%	40%
all MS		
MS + retrob. neuritis		

Lowitzsch et al. (1976) found the following normal and abnormal P100 latency values:

normals	105 ± 4 ms
proven MS	134 ± 36 ms
probable MS	119 ± 20 ms
possible MS	105 ± 4 ms

An extensive review about clinico-VEP correlations has recently been published by Kayed et al. (1978).

Latency asymmetries are found even more often than increased latencies (Collins et al. 1978). Latency asymmetries > 6 to 7 ms are considered to be abnormal (Lehmann and Mir 1976; Kjaer 1980a).

In MS the VEP abnormalities usually run parallel with the visual acuity disturbances; however, even clinically silent abnormalities can be detected. In case of clinical improvement the VEP improves also. However, the SSEP, once found to be abnormal, does not show much recovery in clinical improvement phases (Matthews and Small 1979; Diener and Scheibler 1980).
In summary VEP is an important tool in MS diagnosis. It can detect clinically silent optic system involvement, this way helping to diagnose the disease in an early phase. The choice of the VEP stimulus parameters and of additional tests (SSEP, BAEP) should be considered carefully.
In case of suspected MS it can be advised to apply first pattern reversal stimulation, since this is probably the most appropriate method.
If no abnormalities can be detected with pattern reversal, other stimulation methods (full field of small spot flash) should be considered.

8.7. Uremia

Monitoring studies evaluating the central nervous system condition in uremic patients also include VEP testing. Teschan (1975) found increased latencies. Also Lewis et al. (1978) found increased latenties; P100 in hemodialysis patients 126 ms compared with a normal value of 110 ms.
After hemodialysis improvement occurred (to 122 ms). Also, the amplitude was found to be increased. The same results have been published by Hamel et al. (1978) for N60 latencies:

normal	60 ± 11 ms
moderate uremia	66 ± 17 ms
severe uremia	84 ± 23 ms
after dialysis	72 ± 15 ms
after renal transplantation	67 ± 16 ms

In summary VEP can be used to evaluate the central nervous system condition in uremia, in addition to EMG (maximal motor nerve conduction velocity) of the peripheral nerve system.

8.8. Coma

Bergamini and Bergamasco (1967) described that in coma the amplitude may decrease and the early components of the flash response may even disappear. This effect is not easy to explain. Later studies do not often distinguish between the early and late components. The VEP certainly is influenced by coma, but also by anesthetics. The BAEP is rather indifferent to central nervous system depressiva. So to evaluate an intoxication VEP indicates the level of coma and of intoxication, but BAEP is better to detect structural lesions.

In 1972 Walter and Arfel examined the VEP in deep comatose patients. In acute coma the VEP may disappear before the EEG becomes isoelectric . In "coma dépassé" no VEP can be elicited. One should remember that the ERG can persist for a longer time, so ERG contamination artefacts should not erroneously be regarded as a VEP (Arfel 1967). On the other hand, animal experiments showing VEP to be more resistant to anoxia and improving earlier than EEG does (Horsmann and Kleihues 1973).

Trojaborg and Jørgensen (1973) examined 50 comatose patients with isoelectric EEGs. VEP and SSEP were performed. In 19 patients brain stem reflexes were positive; in all these patients VEPs and SSEPs could be evoked. In 31 patients brain stem reflexes were absent; in 30 of them VEP and SSEP were absent as well. In only 1 out of these 31 patients the SSEP was also absent but the VEP could be evoked. In that particular patient cerebral angiography showed an occlusion of the basilar artery, but hemispheral blood circulation persisted. The patient died nevertheless.

In summary VEP can probably be detected in a few cases of isoelectric EEG. Up till now there are no reports on this finding altering the fatal outcome of isoelectric EEG. Further analysis is necessary to establish the value of VEP testing in cerebral death. The depressive effect of anesthetics and drugs on the VEP should be regarded as a disadvantage. Next to the EEG BAEP is probably the best additional test in the diagnosis of cerebral death.

8.9. Epilepsy

There are a few early papers speaking of "activation" of the VEP during seizures (Bergamini and Bergamasco 1967; Creutzfeldt and Kuhnt 1967; Lucking 1969). There are, however, contradictory results mentioned. Recent papers are scanty; recently Lee et al. (1980) also found increased VEP amplitudes in patients with photoconvulsive seizures.

So far VEP testing has not appeared to be of much value in epilepsy diagnosis.

8.10. Miscellaneous

a. In Parkinson's disease VEP latencies may be delayed. Sometimes improvement occurs by Dopa therapy (Bodis-Wollner and Yahr 1978). Also Gawel et al. (1981) found increased latencies. So in Parkinson's disease there seems to be a subclinically more diffuse involvement of the central nervous system than only the basal ganglia. Maybe this could be regarded as an early sign of a Parkinson dementia syndrome. However, Ehle et al. (1982) could not confirm this.

b. In hypothyreoidea the VEP latency is often found to be increased and the amplitude attenuated (Nishitani and Kooi 1968). On the other hand in hyperthyreoidea the amplitude of all VEP components may be increased (Takahashi and Fujitani 1970).

c. In a number of genetically determined diseases (Friedreich's ataxia, hereditary cerebellar ataxia, familial spastic paraplegia) subclinical VEP abnormalities can be found (Bird and Crill 1981; Pedersen and Trojaborg 1981). The same has been found in Charcot-Marie-Tooth disease (Halliday et al. 1981; Bird and Griep 1981). So there are indications that in many central nervous system degenerative diseases there is a more widespread involvement than can be concluded from the clinical symptoms only.

APPENDIX

Physical aspects of light

1. Methods

1.1. Light and its measurement (photometry)

The visible part of the full range wavelengths of electromagnetic radiation is termed light. This part runs from 400 nm (nanometer = 10^{-9}m) till 700 nm. Adjacent regions are called utraviolet and infrared at smaller and larger wavelengths, respectively.

The visible electromagnetic radiation is not perceived equally at all the wavelengths, which is mainly due to the spectral sensitivity of the photopigments in the retina. Since at high light intensities the cones are governing the visual perception, the quantification of light is performed by using the so-called photopic relative luminous efficiency curve (Fig. 4.22).

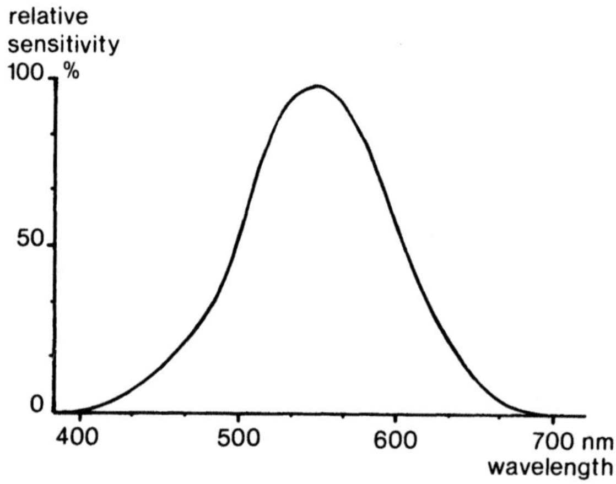

Fig. 4.22 Human photopic relative lumimous efficiency curve: i.e. sensitivity at high luminance levels v.s. wavelength.

The radiaton intensity is measured by the factor obtained from this curve at every wavelenght present. The curve of Fig. was derived from an experiment with 200 human observers and it was adopted as a standard by the "Commission International d'Eclairage" in 1924, hence termed CIE-curve (LeGrand 1968). The physical instruments, i.e. "light meters" are calibrated with this (psychophysical) curve. At low intensities the visual perception is performed by the more sensitive rod-photoreceptors of the retina, which have a different luminous efficiency curve. So, for night vision a correction has to be made to the light meter. In general, however, the VEP measurements are carried out at relatively high light intensities, and moreover the responses are mainly obtained from the central part of the retina, which is dominated by cone photoreceptors. The measurement of the light is a complex and often confusing matter (LeGrand 1968; Zaha 1972). Moreover, it depends on the kind (size) of light source which definitions of light intensity will prevail. The present description is confined to the practically important definitions and measuring units.

a. Point source (e.g. halogen lamp/flash tube, both without diffuser)
The source is considered to be uniform (i.e. radiation equal in all directions) and the unit to express the luminous intensity is termed "candela" (cd) with the dimension "lumen per steradian" ($lm.sr^{-1}$).
b. Extended source (e.g. translucent screen,diffuser)
The source is considered to emit light according to Lambert's law (cosine rule) and the unit to express the luminance is termed "candela per square meter" ($cd.m^{-2}$), which is sometimes called "nit" (nit).

In literature one may sometimes find a different unit of luminance "foot Lambert" (ftL), which is related to the nit as follows:

$$1 \text{ ftL} = 3.426 \text{ nit } (cd.m^{-2})$$

Erroneously, it is often tried to characterize the emission of light by using a light meter at some distance. In fact, what is measured then is the so-called illuminance of the surface by the light source, with the unit "lux" (lx) ($lm.m^{-2}$). Only in case if light is illuminating a diffuse white reflector, with known reflection coefficient (e.g. 0.7), is it possible to deduce the luminance of the reflecting surface (secondary light source) by the formula:

$$\text{luminance} = \text{reflection coefficient} \times \text{illiminance} / \pi$$

$$cd.m^{-2} = r(lx)/\pi$$

A somewhat different problem arises if the light source emits very short duration flashes, so that the above-mentioned light units are no longer sufficient and have to be changed from power units into energy units. So one has to multiply the light units by the duration of the light flash (effective duration). For instance, instead of illuminance: illuminant energy = lux. seconds (lx.s). It will be clear that this measurement cannot be performed with a normal "lux meter". For the purpose of measuring the light output of flash units a so-called "integrating" photometer is needed.

In many publications the flash tube stimulator is characterized by the amount of Watt seconds (Joule) electrically delivered to the tube. Of course this is not an accurate specification, but only an indication of the light output. As a rule of thumb it can be stated that 1 Ws flash will yield a luminant energy of approximately 200 $cd.m^{-2}.s$. When the same tube would be used to illuminate a diffusely reflecting sphere, the flash energy should have to be more than 10 times higher to produce the same luminance.

In case of a small light source it can be imagined that the illuminance of the eye is specifying the effective stimulation of the retina. So, one finds the light stimulation expressed by the amount of lux, or lux.seconds at the cornea, which amount is inversely proportional to the distance between light source and the eye. One should be aware, however, that most light sources may not be considered as a point source because of the relatively large dimensions of the flash tubes and the use of reflectors of various size. Moreover, in general a point source cannot be considered a very adequate stimulus.

The method of calibrating the equipment depends on the kind of light source. It is recommended to place a diffuser (e.g. frosted glass) in front of the source; then the measurement can be performed with a "lux meter", either with a luminance adapter ($cd.m^{-2}$ or ft.L) directly at the diffuser, or with a lux meter at the location of the cornea of the patient.
In the latter case the size of the diffuser should be specified in relation to the distance (visual angle, see below).
A last factor that has to be specified and which determines the efficiency of the light source as a stimulus is the pupil size of the eye of the patient. This may be expressed by the definition of the retinal illuminance by a light source of known luminance (of an extended source):

retinal luminance: luminance x effective pupil area

The unit of retinal illuminance is the "troland" (td) = $cd.m^{-2}.mm^{2}$.
The size of the light source, or of a detail of the scene, is defined by the so-called visual angle (see Fig.4.23).,

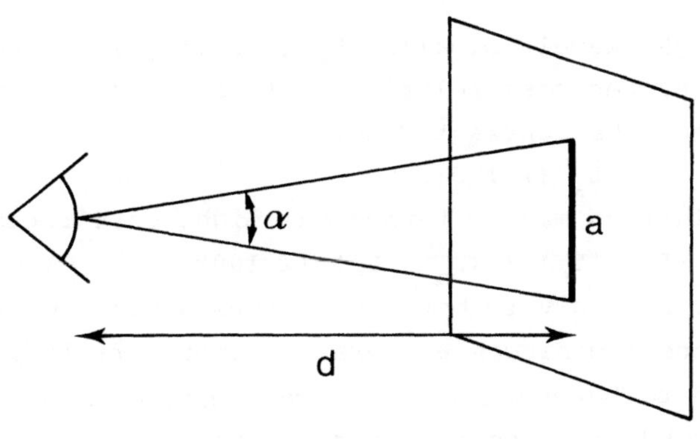

Fig.4.23 The definition of visual angle: α = arctg (a/d)

and is expressed in minutes, or degrees, of arc (e.g. 3°).
The advantage of this measurement is that it is related to the size of the image of the source at the retina.
The relative differences of luminance in a structured stimulus field are defined by the "contrast". Two ways of specifying the contrast (Fig. 4.24) can be used, the former of which is to be preferred.

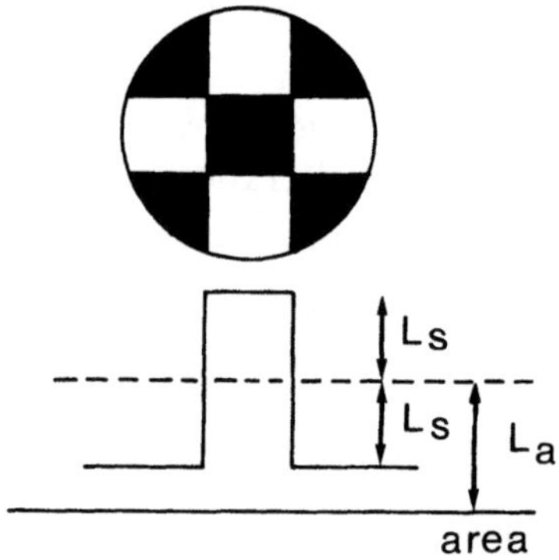

Fig. 4.24 Definition of contrast: increase or decrease of luminance level relative to the average (adaptation) level: C = Ls/La

If the mean light level is called L_a (adapting luminance), and the increase, or decrease relative to this level L_s (stimulus luminance), then the contrast C is:

$$C = L_s / L_a \ (x\ 100\%)$$

The other definition may be termed the light-dark ratio (R), so:

$$R = (L_a + L_s) / (L_a - L_s) \ (x\ 100\%)$$

These definitions apply to both the "simultaneous" contrast, when the various luminances are present in the field at the same time, and to "successive" contrast, when e.g. the adapting light is one condition and the stimulus is added in the next condition and the conditions are not simultaneously present (pattern reversal and pattern onset-offset, respectively).

A few light units and their relations are given below:

luminous intensity	$lm.sr^{-1}$	= cd
luminance	$cd.m^{-2}$	= nit
	1 ft L	= 3.426 $cd.m^{-2}$
	1 asb	= 0.3183 $cd.m^{-2}$
illuminance	$lm.m^{-2}$	= lx
	1 ft.cd	= 10.76 lx

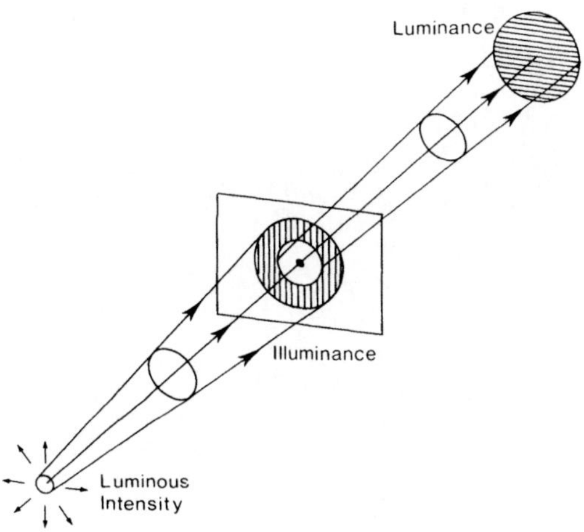

Fig. 4.25 Definition of light measures: luminous intensity = luminous flux per unit of solid angle ($lm.sr^{-1}$ = cd); illuminance = incident luminous flux ($lm.m^{-2}$ = lx); luminance = luminous intensity emitted by extended light source (or reflected by diffuse surface) per unit area of the source ($cd.m^{-2}$ = nit).

1.2. Color (wavelength)

Although the cone photoreceptors determine the VEP predominantly, at low luminances, in the dark, while using large fields, the rod VEP can be isolated (Wooten 1972). If one restricts to photopic luminance levels it appears that the spectral sensitivity of the VEP follows closely the photopic spectral sensitivity curve estimated with behavioural (psychophysical) methods (Fig.4.22). It follows that green light of 550 nm (maximal sensitivity) is approximately 10 times more effective than red light of 650 nm. Therefore, the best choice of the color of a light source will be white, since ideally all the wavelenghts are then equally present and the influence of selective absorption by the optical eye media is greatly reduced. In order to avoid interference of the cone VEP by rod responses a photopic adaptation level of a background field, and a not too large stimulus field (smaller than 30°) should be recommended.

REFERENCES

Abraham, F.A., Melamed,E. and Lavy, S.: Prognostic value of VEP in occipital blindness following basilar artery occlusion. Appl. Neurophysiol., 38: 126-135 (1975).

Allison, T., Matsumiya, Y., Goff, G.D. and Goff, W.R.:The scalp topography of human visual evoked potentials. Electroenceph. clin. Neurophysiol., 42: 185-197 (1977).

Aminoff, M.J. and Ochs, A.L.: Pattern-onset visual evoked potentials in suspected multiple sclerosis. J.Neurol. Neurosurg. Psychiat., 44: 608-614 (1981).

Arfel, G.: Stimulations visuelles et silence cérébral. Electroenceph. clin. Neurophysiol. 23: 172-175 (1967).

Baker, J.B., Larson, S.J., Sances, A. and White, P.T.: Evoked potentials as an aid to the diagnosis of multiple sclerosis. Neurol. (Minneap.), 18: 286 (1968).

Barnet,A.B., Friedman, S.L., Weiss, I.P., Ohlrich, E.S., Shanks, B. and Lodge, A.: VEP development in infancy and early childhood; a longitudinal study. Electroenceph. clin. Neurophysiol., 49: 476-489 (1980).

Bartl, G., Van Lith, G.H.M. and Van Marle, G.W.: Cortical potentials evoked by a TV pattern reversal stimulus varying check sizes and position of the stimulus field. Ophthal. Res., 10: 295-301 (1978).

Begleiter, H., Porjesz, B. and Gross, M.M.: Cortical evoked potentials and psychopathology. Arch. gen. Psychiat., 17: 755-758 (1967).

Bennett, J.R., MacDonald, J.S., Drance, S.M. and Uenoyama, K.: Some statistical properties of the visual evoked potential in man and their application as a criterion of normality. IEEE Trans. Biomed. Engng. BME 18, 1: 23-34 (1971).

Bergamini, L. and Bergamasco, B.: Cortical evoked potentials in man. Thomas, Springfield, p.116 (1967).

Bird, T.D. and Crill, W.E.: Pattern-reversal visual evoked potentials in the hereditary ataxias and spinal degenerations. Ann. Neurol, 9: 243-250 (1981).

Bird, T.D. and Griep, E.: Pattern-reversal visual evoked potentials. Studies in Charcot-Marie-Tooth hereditary neuropathy. Arch. Neurol., 38: 739-741 (1981).

Blom, J.L., Barth, P.G. and Visser, S.L.: The visual evoked potential in the first six years of life. Electroenceph. clin. Neurophysiol., 48: 395-405 (1980).

Blumhardt, L.D., Barrett, G. and Halliday, A.M.: The asymmetrical visual evoked potential to pattern reversal in one half field and its significance for the analysis of visual field defects. Brit. J. Ophthal., 61: 454-461 (1977).

Bodis-Wollner, I. and Yahr, M.D.: Measurements of visual evoked potentials in Parkinson's disease. Brain, 101: 661-671 (1978).

Bostem, F., Rousseau, J.C., Degossely, M. and Dongier, M.: Psychopathological correlations of the non-specific portion of visual and auditory evoked potentials and the associated contingent negative variation. In: W.A. Cobb and C. Morocutti (Eds), The evoked potentials. Electroenceph. clin. Neurophysiol., suppl. 26: 131-138 (1967).

Braakhuis, W.J.M. and Thijssen, J.M.: Standardized electro-ophthalmology using selfcalibrating equipment. Docum. Ophthal. Proc. Series, 15: 317-318 (1978).

Brindley, G.S.: Physiology of the retina and the visual pathway. Edward Arnold Ltd., London (1970).

Brudet-Wickel, C.L.M., Van Lith, G.H.M. and Graniewski-Wijnands, H.S.: Drusen of the optic disc and occipital transient pattern responses. Docum. Ophthal., 50: 243-248 (1981).

Brudet-Wickel, C. and Van Lith, G.: Electrophysiology of anterior ischemic optic neuropathy (AION). Ophthal. in press (1983).

Camacho, L.M., Wenzel, W. und Aschoff, J.: Klinische Anwendung der visuell evozierten Potentiale zur Untersuchung von chiasmatischen und postchiastmatischen Läsionen. Arch. Psychiat. Nervenkr., 230: 243-256 (1981).

Celesia, G.G. and Daly, R.F.: Effects of aging on the visual evoked responses. Arch. Neurol. (Chic.), 34: 403-407 (1977).

Chelva, E. and Van Lith, G.H.M.: Influence of defocusing, absorbtion and scatter on evoked potentials to contrast and contour related stimuli. Docum. Ophthal. Proc. Series, 31: 337-344 (1982).

Cigánek, L.: The EEG response (evoked potential) to light stimulus in man. Electroenceph. clin. Neurophysiol., 13: 165-172 (1961).

Cobb, W.A. and Dawson, G.D.: The latency and form in man of the occipital potentials evoked by bright flashes. J. Physiol. (Lond.) 152: 108-121 (1960).

Coger, R.W., Dymond, A.M., Serafetinides, E.A.,Lowenstam, I. and Pearson, D.: Alcoholism: averaged visual evoked response amplitude - intensity slope and symmetry in withdrawal. Biol. Psychiat., 11: 435-443 (1976).

Collins, D.W.K., Black, J.L. and Mastaglia, F.L.: Pattern reversal visual evoked potential; method of analysis and results in multiple sclerosis. J. Neurol. Sci., 36: 83-95 (1978).

Conners, C.K.: Cortical visual evoked response in children with learning disorders. Psychophysiol., 7: 418-428 (1971).

Cracco, R.Q. and Cracco, J.B.: Visual evoked potential in man: early oscillatory potentials. Electroenceph. clin. Neurophysiol., 45: 731-739 (1978).

Creutzfeldt, O.D. and Kuhnt, U.: The visual evoked potential. Physiological, developmental and clinical aspects. In: W.Cobb and C. Morocutti (Eds.), The evoked potential. Electroenceph. clin. Neurophysiol., suppl. 26: 29-41 (1967).

Crevits, L. and Van Lith, G.: Component analysis of pattern evoked occipital potentials in hemianopic patients. Ophthal. in press (1983).

Crighel, E., Arseni, C. and Matei, M.: Visual evoked responses and their dispersion pattern in frontal lobe tumours. Eur. Neurol., 14: 17-28 (1976).

Diener, H.C. and Scheibler, H.: Follow up studies of visual potentials in multiple sclerosis, evoked by checkerboard and foveal stimulation. Electroenceph. clin. Neurophysiol., 49: 490-496 (1980).

Donchin, E., Callaway, E., Cooper, R., Desmedt, J.E., Goff, W.R., Hillyard, S.A. and Sutton, S.: Publication criteria for studies of evoked potentials (EP) in man. Report of a committee. In: J.E. Desmedt (Ed). Progress in Clinical Neurophysiol., vol. I, Karger, Basel: 1-11 (1977).

Dorfman, L.J. and Cummins, K.L.: Pattern reversal visual evoked potentials in late-onset progressive multiple sclerosis. Electromyogr. clin. Neurophysiol., 19: 89-94 (1979).

Dustman, R.E. and Beck, E.C.: The effects of maturation and aging on the wave form of visually evoked potentials. Electroenceph. clin. Neurophysiol., 26: 2-11 (1969).

Dustman, R.E., Snyder, E.W., Callner, D.A. and Beck, E.C.: The evoked response as a measure of cerebral dysfunction. In:H. Begleiter (Ed.) Evoked brain potentials and behavior. Plenum Press, New York and London : 321-363 (1979).

Duwaer, A.L. and Spekreyse, H.: Latency of luminance and contrast evòked potentials in multiple sclerosis patients. Electro-enceph. clin. Neurophysiol., 45: 244-258 (1978).

Ehle, A.L., Stewart, R.M., Lellelid, N.E. and Leventhal, N.A.: Normal checkerboard pattern reversal evoked potentials in Parkinsonism. Electroenceph. clin. Neurophysiol., 54: 336-338 (1982).

Ellenberger, C.K., Petro, D.J. and Ziegler, S.B.: The visually evoked potentials in Huntington disease. Neurology (Minneap.) 28: 95-97 (1978).

Ellingson, R.J.: Cortical electrical responses to visual stimulation in the human infant. Electroenceph. clin. Neurophysiol., 12: 663-677 (1960).

Feinsod, M., Abramsky, O. and Auerbach, E.: Electrophysiological examinations of the visual system in multiple sclerosis. J. neurol. Sci., 20: 161-175 (1973).

Floris, V., Morocutti, C., Amabile, G., Bernardi, G., Rizzo, P. A. and Vasconetto, C.: Recovery cycle of the visual evoked potentials in normal and schizophrenic subjects. In: W. Cobb and C. Morocutti (Eds). The evoked potentials. Electroenceph. clin. Neurophysiol., suppl. 26: 74-81 (1967).

Gambi, D., Rossini, P.M, Albertini, G., Sollazzo, D., Torrioli, M.G. and Polidori, G.C.: Follow-up of visual evoked potential in full-term and pre-term control newborns and in subjects who suffered from perinatal respiratory distress. Electroenceph. clin. Neurophysiol., 48: 509-516 (1980).

Gastaut, H., and Régis, H.: Visually-evoked potentials recorded transcranially in man. In: D.L. Proctor and W.R. Adey (Eds.) The analysis of central nervous system and cardiovascular data using computer methods. 1964 symposium, NASA, Washington (1965).

Gastaut, H., Régis, H., Lyagoubi, S., Mano, T. and Simon, L.: Comparison of the potentials recorded from the occipital, temporal, and central regions of the human scalp evoked by visual, auditory and somato-sensory stimuli. In: W. Cobb and C. Morocutti (Eds.) The evoked potentials. Electroenceph. clin. Neurophysiol., suppl. 26: 19-28 (1967).

Gawel, M.J., Das, P., Vincent, S., Rose, F.C.: Visual and auditory evoked responses in patients with Parkinson's disease. J. Neuro Neurosurg. Psychiat., 44: 227-232 (1981).

Gliddon, J.B., Busk, J. and Galbraith,G.C.: Visual evoked responses as a function of light intensity in Down's syndrome and non-retarded subjects. Psychophysiol., 12: 416-422 (1975).

Goff, W.R., Allison, T. and Vaughan, H.G.: The functional neuroanatomy of event related potentials. In: E. Callaway, P. Tueting and S.H. Koslow (Eds) Event related brain potentials in man. Academic Press, New York : 1-91 (1979).

Graziani, L.J. and Weitzman, E.D.: Sensory evoked responses in the neonatal period and their application. In: A. Rémond (Ed.) Handbook of EEG and clin. Neurophysiol., Elsevier, Amsterdam, vol. 15b : 73-88 (1972).

Green, J.B. and Walcoff, M.R.: Evoked potentials in multiple sclerosis. Arch. Neurol., 39: 696-697 (1982).

Halliday, A.M.: New applications of visual evoked potentials in man. Jap. J. EEG EMG. suppl. Didactic Lectures 10th ICECN, Kyoto: 117-124 (1981).

Halliday, A.M. and Michael, W.F.: Changes in pattern evoked responses in man associated with the vertical and horizontal meridians of the visual field. J. Physiol., 208: 499-513 (1970).

Halliday, A.M., McDonald, W.I. and Mushin, J.: Delayed visual evoked response in optic neuritis. Lancet, i: 982-985 (1972).

Halliday, A.M., Carroll, W.M. and Jones, S.J.: Visual and somatosensory evoked potential studies in Charcot-Marie-Tooth disease. Electroenceph. clin. Neurophysiol., 52: s84 (1981).

Hamel, B., Bourne, J.R., Ward, J.W. and Teschan,P.E.: Visually evoked cortical potentials in renal failure. Transient potentials. Electroenceph. clin. Neurophysiol., 44: 606-616 (1978).

Harden, A., Pampiglione, G. and Picton-Robinson, N.: Electroretinogram and visual evoked response in a form of 'neuronal lipidosis' with diagnostic EEG features. J. Neurol. Neurosurg. Psychiat., 36: 61-67 (1973).

Harding, G.F. and Rubinstein, M.P.: The scalp topography of the human visually evoked subcortical potential. Invest. Ophthal. Vis. Sci., 19: 318-321 (1981).

Harter, M.R. and White, C.T.: Effects of contour sharpness and check size on visually evoked cortical potentials. Vis. Res. 8: 701-711 (1968).

Hennerici, H., Wenzel, D. and Freund, H.J.: The comparison of small-size rectangle and checkerboard stimulation for the evaluation of delayed visual evoked responses in patients suspected of multiple sclerosis. Brain, 100: 119-136 (1977).

Heninger, G. and Speck, L.B.: Visual evoked responses and mental status of schizophrenics. Arch. gen. Psychiat., 15: 419-426 (1966).

Hossmann, K.A. and Kleihues, P.: Reversibility of ischemic brain damage. Arch. Neurol. (Chic.), 29: 375-384 (1973).

Hrbek, A., Karlberg, P., Kjellmer, I., Olsson, T. and Riha, M.: Clinical application of evoked electroencephalographic responses in newborn infants. I. Perinatal asphyxia. Develop. Med. Child. Neurol., 19: 34-44 (1977).

Hrbek, A., Karlberg,P., Kjellmer, I., Olsson, T. and Riha, M.: Clinical application of evoked EEG responses in newborn infants. II. Idiopathic respiratory distress syndrome. Develop. Med. Child. Neurol., 20: 619-626 (1978).

Huber, C. and Wagner, T.:Electrophysiological evidence for glaucomatous lesions in the optic nerve. Ophthal. Res., 10: 22-29 (1978).

Ikeda, H., Tremain, K.E. and Sanders, M.D.: Neuropsychological investigation in optic nerve disease: combined assessment of the visual evoked response and electroretinogram. Brit. J. Ophthal., 62: 227-239 (1978).

Jackson, S.A. and Barber, C.: The effect of temporal stimulus parameters upon the VEP. In: C. Barber (Ed.) Evoked potentials Univ. Park Press, Baltimore: 175-181 (1980).

Jeffreys, D.A. and Axford, J.G.: Source locations of pattern-specific components of the human visual evoked potentials. I. Component of striate cortical origin. Exp. Brain. Res., 16: 1-21 (1972a).

Jeffreys, D.A. and Axford, J.G.: Source locations of pattern-specific components of the human visual evoked potentials. II. Component of extrastriate cortical origin. Exp. Brain. Res., 16: 22-40 (1972b).

Jonkman, E.J.: The average cortical response to photic stimulation. A study in normal subjects and neurological patients. Thesis, Amsterdam GU, pp160 (1967).

Kayed, K., Røsjø, Ø, and Käss, B.: Practical application of patterned visual evoked responses in multiple sclerosis. Acta neurol. Scand., 57: 317-324 (1978).

Kjaer, M.: Visual evoked potentials in normal subjects and patients with multiple sclerosis. Acta neurol. Scand., 62: 1-13 (1980a).

Kjaer, M.: Brain stem auditory and visual evoked potentials in multiple sclerosis. Acta neurol. Scand., 62: 14-19 (1980b).

Kjaer, M.: The value of brain stem auditory, visual and somato-sensory evoked potentials and blink reflexes in the diagnosis of multiple sclerosis. Acta neurol., Scand., 62: 220-236, (1980c).

Kojima, M. and Zrenner, E.: Determination of local thresholds in the visual field by recording the scotopic visually evoked cortical potential in man. Ophthal. Res., 12: 1-8 (1980).

Laget, P., Salbreux, R., Raimbault, J., D'Allest, A.M. and Mariani, J.: Relationship between changes in somaesthetic evoked potentials and EEG findings in children with hemiplegia. Develop. Med. Child Neurol., 18: 620-631 (1976).

Laget, P., Flores-Guevara, R., D'Allest, A.M., Ostre, C., Raimbault, J. et Mariani, J.:La maturation des potentiels évoqués visuels chez l'enfant normal. Electroenceph. clin. Neurophysiol., 43: 732-744 (1977).

Larson, S.J., Sances, A. and Baker, J.B.: Evoked potentials in patients with stroke. Circulation, 33, suppl. 2: 15-19 (1966).

Laurian, S., Lobrinus, S., Wertheimer, J. and Gaillard, J.M.: Evoked responses in dementia and the significance of the vertex potential. Electroenceph. clin. Neurophysiol., 43: 525 (1977).

Lee, R.G. and Blair, D.G.: Evolution of EEG and visual evoked response changes in Jakob Creutzfeldt disease. Electroenceph. clin. Neurophysiol., 35: 133-142 (1973).

Lee, S.I., Messenheimer, J.A., Wilkinson, E.C., Brickley Jr., J.J. and Johnson, R.N.: Visual evoked potentials to stimulus trains: normative data and application to photosensitive seizures. Electroenceph. clin. Neurophysiol., 48: 387-394 (1980).

LeGrand, Y.: Light,color and vision. Chapman and Hall, London (1968).

Lehmann, D. und Mir, Z.: Methodik und Auswertung visuell evozierter EEG-Potentiale bei Verdacht auf Multiple Sklerose. J.Neurol., 213: 97-103 (1976).

Lehmann, D. und Soukos, I.: Visuell evozierte Potentiale und Hirnstamm-klick-Potentiale in der Frühdiagnose der Multiplen Sklerose: Statistik. Nervenarzt, 53: 327-332 (1982).

Lesèvre, N.: Topographical analysis of the pattern evoked response (PER): its application to the study of macular and peripheral vision in normal people and in some pathological cases. Docum. Ophthal. 10: 87-102 (1976).

Lewis, E.G., Dustman, R.E. and Beck, E.C.: Visual and somatosensory evoked potentials characteristics of patients undergoing hemodialysis and kidney transplantation. Electroenceph. clin. Neurophysiol., 44: 223-231 (1978).

Liberson,W.T. and Scott, D.: Evoked potentials in hemiplegic patients. Electroenceph. clin. Neurophysiol, 17 : 451 (1964).

Lowitzsch, K., Kuhnt, U., Sackmann, Ch., Maurer, K., Hopf, H.C., Schott, D. and Thäter, K.: Visual pattern evoked responses and blink reflexes in assessment of MS diagnosis (a clinical study of 135 multiple sclerosis patients). J. Neurol., 213: 17-32 (1976).

Lucking, C.H.: Variability of visual evoked responses in epileptic patients. Electroenceph. clin. Neurophysiol., 27: 702 (1969).

Marcus, M.M.: Visual evoked potentials to flash and pattern in normal and high risk infants. In: J.E. Desmedt (Ed.) Visual evoked potentials in man; new developments. Clarendon Press, Oxford: 490-499 (1977).

Maitland, C.G., Aminoff, M.J., Kennard, C. and Hoyt, W.F.: Evoked potentials in the evaluation of visual field defects due to chiasmal or retrochiasmal lesions. Neurology (NY), 32: 986-991 (1982).

Matthews, W.B. and Small, D.G.: Serial recording of visual and somatosensory evoked potentials in multiple sclerosis. J. neurol. Sci., 40: 11-21 (1979).

Matthews, W.B., Small, D.G., Small, M. and Pountney,E.: Pattern reversal evoked visual potential in the diagnosis of multiple sclerosis. J.Neurol. Neurosurg. Psychiat., 40: 1009-1014 (1977)

Mellor, D.H. and Fielder, A.R.: Dissociated visual development: electrodiagnostic studies in infants who are "slow to see", Develop. Med. Child Neurol., 22: 327-335 (1980).

Müller-Jensen, A., Zschocke, S. and Dannheim, F.: VER analysis of the chiasmal syndrome. Neurology (NY), 225: 33-40 (1981).

Neetens, A., Hendrata, Y. and Van Rompaey, J.: Pattern and flash visual evoked responses in multiple sclerosis. J. Neurol., 220: 113-124 (1979).

Nishitani, N. and Kooi, K.A.: Cerebral evoked responses in hypothyroidism. Electroenceph. clin. Neurophysiol. 24: 554-560 (1968).

Njiokiktjien, Ch.J., Visser, S.L. and De Rijke, W.: EEG and visual evoked responses in children with learning disorders. Neuropädiatrie, 8: 134-147 (1977).

Oepen, G., Brauner, C., Doerr, M. and Thöden, U.: Visual evoked potentials (VEP) elicited by checkerboard versus foveal stimulation in multiple sclerosis. A clinical study in 235 patients. Arch. psychiat. Nervenkr., 229: 305-313 (1981).

Oepen, G., Doerr, M. and Thöden, U.: Huntington's disease: alterations of visual and somatosensory cortical evoked potentials in patients and affspring. In: J.Courjon, F. Mauguiere and M.Revol (Eds.) Clinical applications of evoked potentials in neurology. Raven Press, New York: 141-147 (1982).

Oosterhuis, H.J.G.H., Ponsen, L., Jonkman, E.J. and Magnus, O.: The average visual response in patients with cerebrovascular disease. Electroenceph. clin. Neurophysiol., 27: 23-34 (1969).

Pampiglione, G.: Electrical activity of the brain in some genetically determined neurometabolic diseases. In: H. van Duijn, D.N.J. Donker and A.C. van Huffelen (Eds.) Current concepts in clinical neurophysiology. Trio, Den Haag: 23-30 (1977).

Pedersen, L. and Trojaborg, W.: Visual, auditory and somatosensory pathway involvement in hereditary cerebellar ataxia, Friedreich's ataxia and familial spastic paraplegia. Electroenceph. clin. Neurophysiol., 52: 283-297 (1981).

Persson, H.E. and Wanger, P.: Pattern-reversal electroretinograms in squint amblyopia, artificial anisometropia and simulated eccentric fixation. Acta Ophthalm., 60: 123-132 (1982).

Poortinga, K., Njiokiktjien, Ch.J. and Visser, S.L.: Correlations between visual evoked potentials of 5-year-old children and their neonatal neurological condition. In: J. Courjon, F. Mauguiere and M. Revol (Eds.) Clinical applications of evoked potentials in neurology. Raven Press, New York: 81-87 (1982).

Porjesz, B. and Begleiter, H.: Visual evoked potentials and brain dysfunction in chronic alcoholics. In: H. Begleiter (Ed.) Evoked brain potentials and behavior. Plenum Press, New York and London : 277-302 (1979).

Posthuma, J. and Visser, S.L.: Visual evoked potentials and alcohol-induced brain damage. In: J. Courjon, F.Mauguiere and M. Revol (Eds.) Clinical applications of evoked potentials in neurology. Raven Press, New York: 149-155 (1982).

Prechtl, H.F.R.: The neurological examination of the full-term newborn infant. Clinics in developmental medicine, vol. 63, 2nd ed. Heinemann, London, pp 68 (1977).

Preston, M.S., Guthrie, J.T. and Childs, B.: Visual evoked responses (VERs) in normal and disabled readers. Psychophysiol., 11 : 452-457 (1974).

Regan, D. and Richard, W.: Independence of evoked potentials and apparent size. Vis. Res., 11 : 679-684 (1971).

Rémond, A. et Lesèvre, N.: Distribution topographique et potentiels évoqués visuels occipitaux chex l'homme normal. Rev. Neurol., 112: 317-330 (1965).

Richey, E.T., Kooi, K.A. and Tourtellotte, W.W.: Visually evoked responses in multiple sclerosis. J.Neurol. Neurosurg. Psychiat., 34: 275-280 (1971).

Rietveld, W.J., Tordoir, W.E.M., Hagenouw, J.R.B., Lubbers, J.A. and Spoor, T.A.C.: Visual evoked responses to blank and to checkerboard patterned flashes. Acta physiol. pharmacol. neerl., 14: 259-285 (1967).

Rossini, P.M., Caltagirone, C., David, P. and Macchi, G. : Jakob-Creutzfeldt disease: analysis of EEG and evoked potentials under basal conditions and neuroactive drugs. Eur. Neurol., 18: 269-279 (1979).

Rowe III, M.J.: The clinical utility of half-field pattern reversal visual evoked potential testing. Electroenceph. clin. Neurophysiol., 53: 73-77 (1982).

Salamy, J.G., Wright, J.R. and Faillace, L.A.: Changes in average evoked responses during abstention in chronic alcoholics. J.Nerv. Ment. Dis., 168: 19-25 (1980).

Schneider, J.: Tumeurs cérébrales et potentiels évoqués. Rev. Neurol., 118: 443-458 (1968).

Shagass, C.: Evoked brain potentials in psychiatry. Plenum Press, New York, pp 274 (1972).

Shagass, C. and Schwartz, M.: Visual cerebral evoked response characteristics in a psychiatric population. Amer. J. Psychiat 121: 979-987 (1965).

Shahrokhi, F., Chiappa, K.H. and Young, R.R.: Pattern shift visual evoked responses. Two hundred patients with optic neuritis and/or multiple sclerosis. Arch. Neurol. (Chic.), 35: 65-71 (1978).

Shearer, D.E. and Dustman, R.E.: The pattern reversal evoked potential. The need for laboratory norms. Amer. J. EEG Technol., 20: 185-200 (1980).

Siegfried, J.B. and Lukas, J.: Early wavelets in the VECP. Invest. Ophthal. Vis. Sci., 20: 125-129 (1981).

Sokol, S.: Visually evoked potentials: theory, techniques and clinical applications. Surv. Ophthal., 21 : 18-44 (1976).

Sokol, S., Moskowitz, A. and Towle, V.L.: Age-related changes in the latency of the visual evoked potential: influence of check size. Electroenceph. clin. Neurophysiol., 51: 559-562 (1981).

Speck, L.B., Dim, B. and Mercer, M.: Visual evoked responses of psychiatric patients. Arch. gen. psychiat., 15: 59-63 (1966).

Spekreyse, H.: Analysis of EEG responses in man, evoked by sinewave modulated light. Thesis, University of Amsterdam. Junk Publ., The Hague, pp 163 (1966).

Spekreyse, H., Van der Tweel, L.H. and Zuidema,T.: Contrast evoked responses in man. Vis. Res., 13: 1577-1601 (1973).

Symann-Louett, N., Gascon, G.G., Matsumiya, Y. and Lombroso, C.T.: Wave form difference in visual evoked responses between normal and reading disabled children. Neurology (Minneap.), 27: 156-159 (1977).

Takahashi, K. and Fujitani, Y.: Somatosensory and visual evoked potentials in hyperthyroidism. Electroenceph. clin. Neurophysiol., 29: 551-556 (1970).

Teschan, P.: EEG and other neurophysiological abnormalities in uremia. Kidney internat. Nephrol. : 210-216 (1975).

Thorwirth, V.: Visual evoked potentials in cerebrovascular disease. Electroenceph. clin. Neurophysiol., 52: S31 (1981).

Thijssen, J.M., Braakhuis, W., Pinckers, A. and Van Lith, G.: Standardized electro-ophthalmography. Proc. 170th meeting Dutch Ophthal. Soc.,Junk Publ., The Hague: 185 (1976).

Trojaborg, W. and Jørgensen, E.O.: Evoked cortical potentials in patients with 'isoelectric' EEGs. Electroenceph. clin. Neurophysiol., 35: 301-309 (1973).

Van Lith, G.H.: Difficulties in diagnosing Leber's optic atrophy. Docum. Ophthal., 48: 255-259 (1980a).

Van Lith, G.H.M.: Electro-ophthalmology. In: I.C. Michaelson (Ed.). Textbook of the fundus of the eye. Churchill Livingstone, 3rd ed.: 23-24 (1980b).

Van Lith, G.H.M.: Application of electro-ophthalmology. In: I.C. Michaelson (Ed.) Textbook of the fundus of the eye. Churchill Livingstone, 3rd ed.: 627-655 (1980c).

Van Lith, G.H.M. and Henkes, H.E.: The relationship between ERG and VER. Ophthal. Res., 1: 40-47 (1970).

Van Lith, G.H.M. and Henkes, H.E.: Low vision in alcohol abuse. Docum. Ophthal., 46: 333-338 (1979).

Van Lith, G.H.M. and Vijfvinkel-Bruinenga, S.: Optic neuropathy due to alcohol abuse and evoked cortical potentials. Docum. Ophthal. Proc. Series, 15: 221-225 (1978).

Van Lith, G.H.M., Van Marle, G.W. and Van Dok-Mak, G.T.M.: Variation in latency times of visually evoked cortical potentials. Brit. J. Ophthal., 62: 220-222 (1978).

Van Lith, G.H.M., Henkes, H.E. and Vijfvinkel-Bruinenga, S.M.: Asymmetric pattern evoked responses and stimulus parameters. Docum. Ophthal. Proc. Series, 23: 249-253 (1981).

Van Lith, G.H.M., Vijfvinkel-Bruinenga, S. and Graniewski-Wijnands, H.: Pattern evoked cortical potentials and compressive lesions along the visual pathways. Docum. Ophthal., 52: 347-355 (1982).

Vaughan, H.G., Katzman, R. and Taylor, J.: Alterations of visual evoked response in the presence of homonymous visual defects. Electroenceph. clin. Neurophysiol., 15: 737-746 (1963).

Visser, S.L., Stam, F.C., Van Tilburg, W., Op den Velde, W., Blom, J.L. and De Rijke, W.: Visual evoked response in senile and presenile dementia. Electroenceph. clin. Neurophysiol., 40: 385-392 (1976).

Visser, S.L., Poortinga, K. and Njiokiktjien, Ch.J.: Correlation between EEG abnormalities at the age of 5 and a non-optimal neurological condition at birth. Electroenceph. clin. Neurophysiol., 52: S5-S6 (1981).

Visser, S.L., Njiokiktjien, Ch.J. and De Rijke, W.: Neurological condition at birth in relation to the electroencephalogram (EEG) and visual evoked potential (VEP) at the age of 5. Electroenceph. clin. Neurophysiol., 54: 458-464 (1982).

Walsh, J.C., Garrick, R., Cameron, J. and McLeod, J.G.: Evoked potential changes in clinically definite multiple sclerosis; a two year follow up study. J.Neurol. Neurosurg. Psychiat., 45: 494-500 (1982).

Walter, S. et Arfel, G.: Responses aux stimulations visuelles dans les états de coma aigu et de coma chronique. Electroenceph. clin. Neurophysiol., 32: 27-41 (1972).

Wilson, W.B. and Keyser, R.B.: Comparison of the pattern and diffuse-light visual evoked responses in definite multiple sclerosis. Arch. Neurol. (Chic.), 37: 30-34 (1980).

Wooten, B.R.: Photopic and scotopic contributions to the human visually evoked cortical potential. Vis. Res., 10: 1647-1660 (1972).

Wijngaarde, R. and Van Lith, G.H.M.: Electrodiagnostics of the tilted disc syndrome. Docum. Ophthal., 50: 365-369 (1981).

Zaha, M.A.: Shedding some needed light on optical measurements. Electronics, 6: 91-96 (1972).

Chapter V

Somatosensory Evoked Potentials

E.J. Colon
F.E. Posthumus Meijjes

1. INTRODUCTION

Somato sensory evoked potentials are potentials which are elicited by means of stimuli on the skin, sensory organs or the sensory nerves. They are derived from the peripheral nerves or the central nervous system.
For the latter type a further subdivision can be made into spinal, brainstem and cortical evoked potentials, according to the structures from which the response derives.
Up until now, the origin of the different components of the somato sensory evoked potentials has been only partially known. However, the use of these potentials in the diagnosis of lesions in the nervous system has proven to be valuable.
The value, especially of the somato sensory evoked cortical potential, in relation to the other evoked potentials, is seen in the fact that the SSEP is clearly composed of two distinct components, of which the first is a stable part produced by the exitation of a very long myelinated pathway. Abnormalities in this myelinated pathway may therefore be found earliest by SSEP.
The later components are influenced by the functional state of the brain. In clinical neurophysiology the SSEP seems especially suitable for the differentiation between diseases in the myelinated system and global changes in the brain itself. There are, as will be demonstrated in this chapter, many other applications. The signal/noise ratio is especially poor in SSEP, which means that a relatively large number of stimuli must be given. Given the increased need for gathering information by means of the SSEP in clinical practice, an increasing interest develops towards methods that might

decrease this number of stimuli to be given to reach an interpretable signal. The TVF (Time Varying Filtering, de Weerd J., 1981) is in this respect one of the methods that seems to be appropriate.

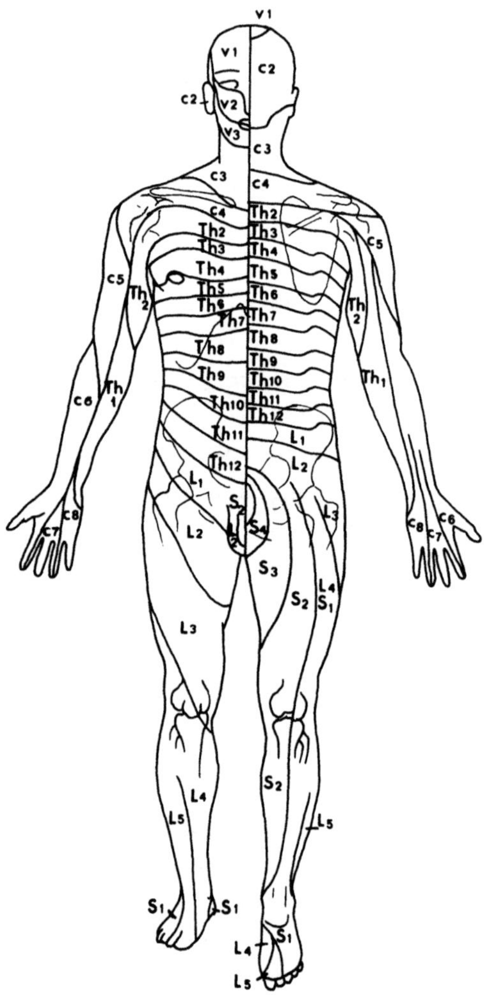

Fig.5.1A Segmental sensory distribution in the human.
Front and back.

2. ANATOMY AND PHYSIOLOGY

2.1 The skin and its sense organs (Fig. 5.1A and 5.1B).

The skin contains different well-known forms of sense organs as well as free nerve endings. Formerly these different sense organs or terminal corpusculus were thought to have specific sensory qualities for touch, heat and pressure. The free nerve endings in the skin function in perceiving noxious stimuli. It is clear, however, from a physiological point of view, that they all transform the various kinds of stimuli, according to whether they are rapid or slow adapting, into electrical signals which travel along the afferent nerve fibres as action potentials.

Apart from the skin, from which all the exteroceptive impulses come, there is also input from sensory organs in the peri-articular tissues and internal organs, giving information about posture and movement (called propriocepsis), and also from sensory organs and free nerve endings in the various visceral organs.

The skin is innervated by cutaneous nerves; most but not all of them are branches from mixed, sensory-motor nerves.

2.2 The peripheral nerve (Fig. 5.1A and 5.1B and Table 5.1).

There are pure sensory nerves, pure motor nerves and mixed nerves. In cutaneous nerves about 50% of the sensory fibres originate from so-called nociceptors; these are specialised receptors that after stimulation give rise to the sensation of pain.

The nerve fibre is formed by the axon of a nerve cell with its enveloping myelin sheath. This myelin sheath is formed by the Schwann's cells. On the other hand, there are also unmyelinated or naked fibres.

The thickest myelinated A-fibres vary from 1 μm tot 20 μm in diameter. Their excitability threshold is low and their conduction velocity is high, up until about 100 m/sec.

(see table 5.1).
The thin unmyelinated <u>C-fibres</u> have a high excitability threshold and their conduction velocity is as low as about 2 m/sec. All the nerve fibres of varying thickness are held together by connective tissue, of which fascicles of nerves fibres in turn are enclosed in a sheat called the perineurium.

1 N. Trigeminus
2 N. auricul. magn.
3 N. cutaneus colli
4 Nn. supraclaviculares
5 Rami ventrales nervor intercost.
6 N. cutaneus brachii rad. (e n. axillare)
7 N. cutaneus brachii uln.
8 Rami laterales nervorum intercost.
9 N. cutaneus antebrachii dors. (e n. radiale)
10 Rami ventr. nervor intercostalium
11 N. cutaneus antebrachii uln.
12 N. cut. antebrachii rad. (e n. musculocut.)
13 R. superfic. ni. rad.
14 R. palm. ni. med.
15 N. med. (Nn. digit. vol.)
16 R. vol. manus ni. uln. (Nn. digit. vol.)
17 R. palm. ni. uln.
18 N. iliohypogastricus (R. cut. lat.)
19 N. ilioinguinalis (Ri. scrotales)
20 N. iliohypogastricus (R. ventr.)
21 N. genitofemoralis (R. femoralis)
22 N. cutaneus femor. later.
23 N. femoralis (Ri. cut. ventr.)
24 N. obturatorius (R. cut.)
25 N. cutaneus surae later.
26 N. saphenus
27 N. peronaeus superfic.
28 N. suralis
29 N. peronaeus prof.
30 N. tibialis (Rr. calcaneares)

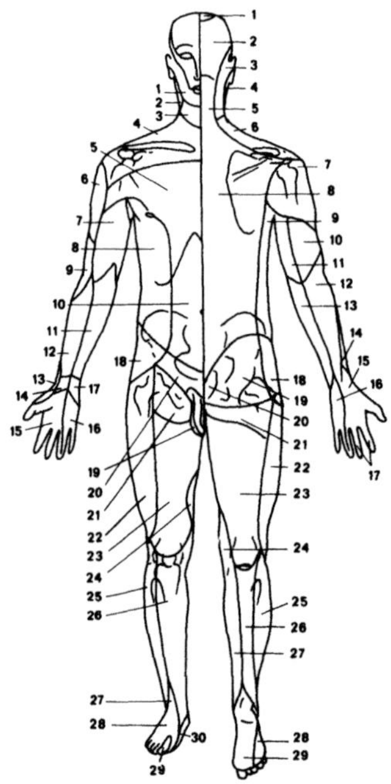

1 N. frontalis (V_1)
2 N. occipitalis major
3 N. occipitalis minor
4 N. auricularis magnus
5 Rami dorsales nervor. spinal. cervic.
6 Nn. supraclaviculares
7 N. cutaneus brachii rad. (e n. axillare)
8 Rami dors. nervor. spin. cervic., thorac., lumb.
9 Rami lat. nervor. intercost.
10 N. cut. brachii dors. (e n. rad.)
11 N. cut. brachii uln.
12 N. cut. antebrachii dors. (e n. radiale)
13 N. cut. antebrachii uln.
14 N. cut. antebrachii rad. (e n. musculocut.)
15 R. superfic. ni. rad. (Nn. digit. dors.)
16 R. dors. manus ni. uln. (Nn. digit. dors.)
17 N. medianus (Nn. digit. vol.)
18 N. iliohypogastricus (R. cut. lat.)
19 Nn. clunium sup.
20 Nn. clunium med.
21 Nn. clunium inf.
22 N. cutaneus femoris lat.
23 N. cutaneus femoris dors.
24 N. obturatorius (R. cut)
25 N. cutaneus surae lat.
26 N. suralis
27 N. saphenus
28 N. plantaris lateralis
29 N. plantaris medialis

Fig. 5.1B Distribution of the sensory nerves over the skin. Front and back.

The impulses or action potentials of each nerve fibre are conducted towards the dorsal ganglion in the dorsal root of the spinal nerve and then forward to the posterior horn in the spinal cord.
All these single fibre action potentials constitute the compound <u>sensory nerve action potential (SNAP)</u>.

Because of the different conduction velocities, the SNAP becomes more dispersed, i.e. of lower amplitude and of longer duration as the distance between the site of stimulation and the site of recording increases.

To measure sensory conduction velocities on peripheral nerves one can use two methods. In the orthodromic method the stimulating electrodes are positioned at the distal end of the limb, where we generally find only sensory nerves. Then the cathode has to be placed proximal, generally on the mixed nerve.

The other possibility is the antidromic way, stimulating proximal and recording distally. In this case the cathode has to be placed distal of the anode. Both methods yield about the same conduction velocity, which for the upper extremity is 50 - 70 m/sec. At the lower limb the mean sensory conduction velocity is 45 - 60 m/sec.

If one stimulates a mixed nerve, for instance the median nerve at the wrist, then the compound action potential contains a volley of antidromically travelling motor nerve action potentials as well.

	A	B and C	C
Function	alpha mechano receptor / delta noci ceptor	autonomic pre- and postganglion	unmyelinated dorsal root fibres warmth nociceptor
Diameter μ	22 - 1	3 - 0.3	1.2 - 0.4
conduction velocity ms	120 - 5	14 - 0.7	2.0 - 0.6
spike duration ms	0.4 - 0,5	1.2 - 2.0	2.0
abs. refractory period ms	0.4 - 1.0	1.2 - 2.0	2.0

Table 5.1 Different properties and parameters of the peripheral nerves.

2.3 Transformation of the information on the segmental level

The dorsal roots (exclusively sensory nerve fibres) contain in their medial part mostly thick myelinated fibres which divide in an ascending branch in the dorsal column and a small descending branch.

The lateral part of the dorsal root is formed mainly by thin and unmyelinated nerve fibres.

For the conduction of impulses giving rise to evoked responses in the more rostral parts of the central nervous system, the dorsal columns are the most important pathways. The conduction velocity of these spinal fibre tracts is 60 m/sec.

In the clinical sensory examination, the sensory qualities of sense of posture, sense of movement and vibration sensation are mainly confined to these dorsal columns.

The information about the site of stimulation and the kind of stimulus, is mainly transported in the spatio-temporal pattern of the volley of impulses that reach the cell groups in the dorsal horn at the segmental level of the spinal cord.

2.4 The spino-cortical pathway (see figure 5.2).

The gnostic information is transported mainly in the dorsal column of the spinal cord. Information from the foot is located in the most median part of this column in the tractus gracialis. Information from the upper extremity is transported over the most lateral part of the dorsal column, the so-called tractus-cuneatus.

When reaching the brainstem, the information is transformed by means of a synapse in the nucleus gracilis and nucleus cuneatus onto the lemniscus medianus and transported to the thalamus on the hetero lateral side. This part of the tract for gnostic information transport is defined as the "second neuron". The "third" neuron is found in the last part of this pathway, when the information is brought from the N. ventro-lateralis of the thalamus (synapse) to the specific sensory cortex.

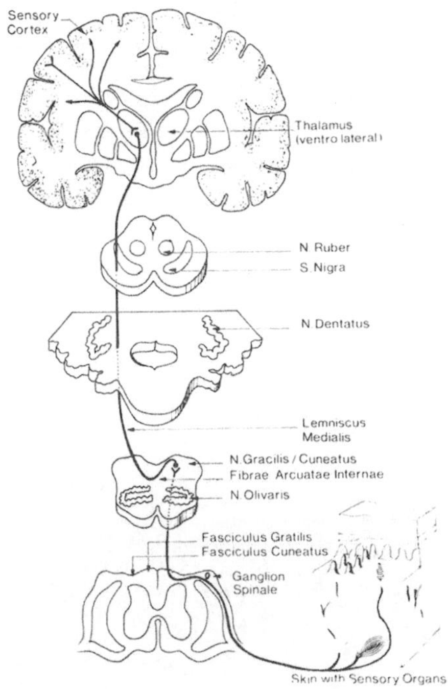

Fig. 5.2 Pathway for the gnostic sensory information from skin to the cerebral cortex.

There is a typical division of body representation over the sensory cortex, which means that for the analysis of the somato sensory evoked response the active electrode must be located above the representing area of the stimulated part of the body. This representation is given in figure 5.3.

2.5 The brain

Figure 5.4 gives a schematic drawing of the way in which information is received and distributed in the brain. The Thalamo-cortical tract ends in the layer IV of the cortex where synapses are found especially with stellate cells. From here the information is transported to the layers II and III and V and VI respectively. From the upper layers, the so-called U-fibres emerge, making contact with adjacent areas.

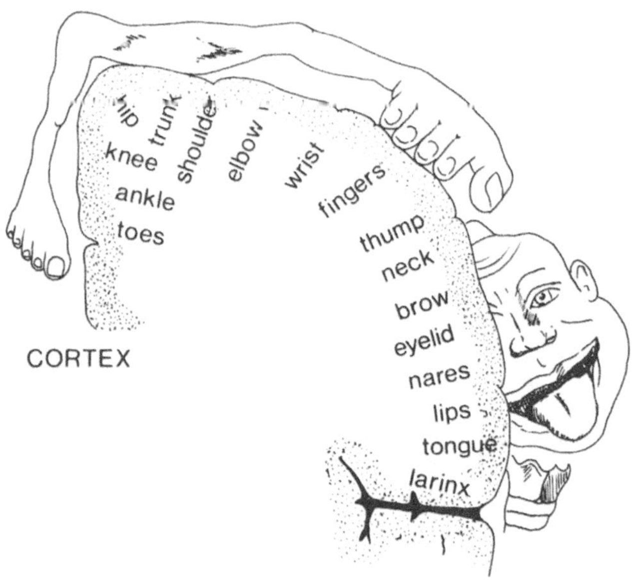

Fig. 5.3 Localisation of the sensory input on the cerebral cortex.

From the lower layers large myelinated fibres emerge to the structures that are relatively far away as, for example, the cortex of the other hemisphere, crossing the corpus callosum.

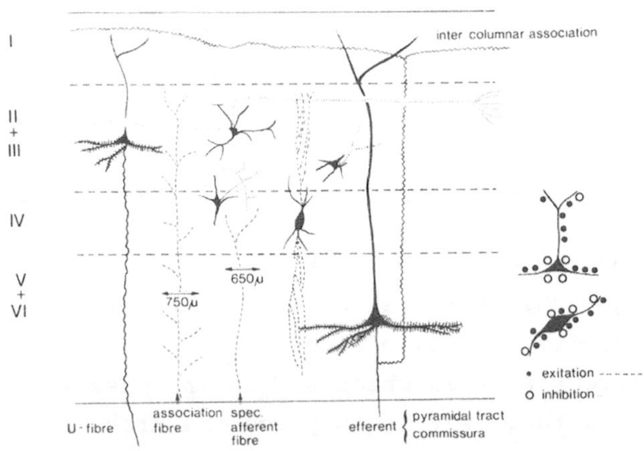

Fig.5.4 Schematic drawing of the cerebral cortex. Information reaches the cortex by association and specific afferent fibres and leaves the cortex by means of efferent and U-fibres.

From the basic work of Hubel and Wiesel (1962) we know that the cortex itself is built up by means of functional columns in which one exclusive function is located. These columns exist out of layers of neurons in such a way that in between different species these layer structures can always be found but are essentially different in their composition.

3. METHOD OF STIMULATION AND RELATED PROBLEMS

3.1 Temperature

The propagation velocity of action potentials in myelinated nerves is substantially influenced by temperature.
The dispersion of action potentials reaching the deriving electrode increases with decreasing temperature, which also gives rise to a broadening of the compound action potentials.
This means that it is only possible to compose normative data for amplitude and latency in populations as well as in individuals in a standard temperature condition.
For clinical use a skin temperature, of at least 30°C (86°F) at the area of stimulation, is needed. The temperature near the nerve is about 3°C (5,4°F) higher.
You must consider the fact that at higher temperature the propagation velocity will also rise.
There are different ways in which we can increase skin temperature. The most suitable way is to place the area of stimulation in warm water or to place an infra-red lamp above this area. One must take care not to burn the skin, especially when using infra-red light. It is advisable always to stay near the patient during warming up. There are commercially available digital thermometers which measure skin temperature quickly and accurately. These thermometers are a must in the EP-room. In every EP experiment the skin temperature over the stimulated nerve must be registrated.

3.2 The stimulus

For somato sensory evoked potential studies we need a stimulus that is clearly defined in time. This means that an electrical stimulus is preferable over tapping or pricking.
Generally, a short electrical stimulus of 0.1 ms (constant current) is applied to a sensory or mixed nerve.
For clinical use a frequency in between ¼ and 10 Hz is generally used for somato sensory stimulation depending on

the clinical application (see later). A higher frequency may give an unpleasant feeling, but will also give changes in latency of components. At low stimulus intensities, only a small part of the myelinated fibres of a nerve will be stimulated. With increasing intensity more and more nerves become active, and the derived compound action potential will grow until a maximum. This maximum is generally reached when the current, in humans, lies between 8 and 15 mA for skin electrodes. The most adequate way for establishing optimal reproducable intensity of the stimulus on the mixed nerve is to give such a current that small twitches are seen in the distal muscle. For pure sensory nerves the amplitude of the distal sensory nerve action potential (SNAP) should be measured and the current should be increased until this potential becomes maximal; one then reaches the supramaximal level.

For example, in studies of the cortical SSEP after stimulation of an index finger, one first analyses the median nerve sensory potential at the wrist. When with increasing intensity there is no increase of this potential, one diminishes the intensity until the moment that the potential also dominishes.

Just above this stimulation level is the optimal intensity (supramaximal), where most nerves are activated and the burden for the patient is relatively minimal.

In this case, one stimulates supramaximal in relation to the distal sensory potential. When stimulating at this supramaximal intensity the maximum voltage of the different components of the SSEP is always reached.

From our own experiments we know that at finger stimulation the latency of N20 reaches its maximum amplitude at a stimulus strength of more than 10 mA, but this value may be higher in persons with a thick finger skin, or fatty tissues. The same holds true for the voltage of the N20 and P200 components (see figure 5.5). In figure 5.5 the effect of the stimulation strength on the amplitude of the different components of the SSEP is given.

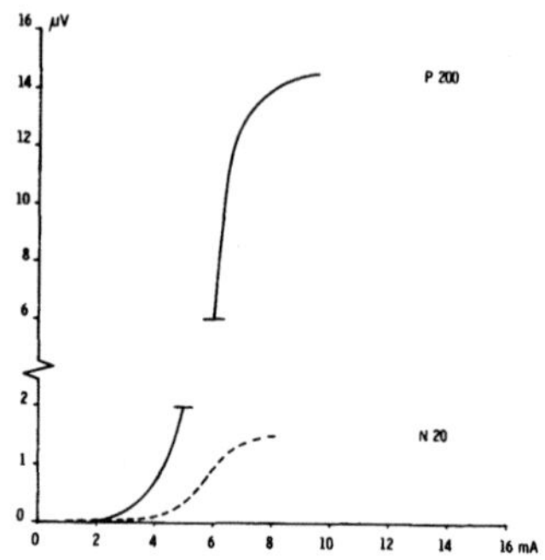

Fig.5.5 Relation between amplitude (in µV) of the cortical N20 and P200 and stimulus strength (in mA) after stimulation of the index finger with ring-electrodes.

The pattern of stimulation may be regular or random.
It seems adequate always to give random stimulation for the non-specific components of the SSEP. Each human input is bound to habituation, which in principle, always appears in repetitive stimulation. This may give rise to changes in amplitude and/or latencies. One can, to some extent, avoid this by using random stimulation.
When performing cortical evoked potential examination the late components have a limiting effect on the time in between which two stimuli may succeed. The recovery phase of the late components is very long (>3 s), which means that the at random stimulis should be given with intervals in between 3 and 5 seconds.
One must consider the possibility of a CNV (Contingent Negative Variation) when using these long time lags in between two stimuli, which is an additional reason for random stimulation.
Alpha rhythm artifacts may also be diminished by random stimulation

The number of stimuli to be given is highly dependent on the signal-to-noise ratio. For distal sensory nerve action potential measurements, a number of 16 stimuli is generally sufficient. For the study of the EP in the neighbourhood of the vertebral column and for the specific cortical reaction, a number of stimuli in between 500 and 1000 may be sufficient. For the study of late components of the SSEP 100 stimuli generally give adequate results.

The electrodes used for the electrical stimulation are given in figure 5.6. The advantage of electrical stimulation is also found in the fact that triggering of these stimuli and the measurement of the intensity of these stimuli is easy.

It is possible, of course, to connect a trigger system to a hammer.In that case, the EP is elicited by means of tapping, which in practice gives the same EP pattern as electrical stimulation. Besides the routine electrical stimulation of sensory or mixed nerves, other methods are used. Mechanical stimulation as tapping, pin-pricking and touch must be mentioned here. Also EP's after thermal and noxious stimuli and joint displacement have been described, but at this moment are not used in a routine clinical setting.

The human somato sensory system can also be driven by stimulus trains, this response can be recognized at frequencies as high as 200 Hz. (Namerow et al. 1974).

3.3 Place of stimulation

There are three ways to chose a place of stimulation, which are highly dependent on the reason why the EP is done.

The first way is segmental stimulation,(performed when one wants to know where abnormalities are located in the spinal

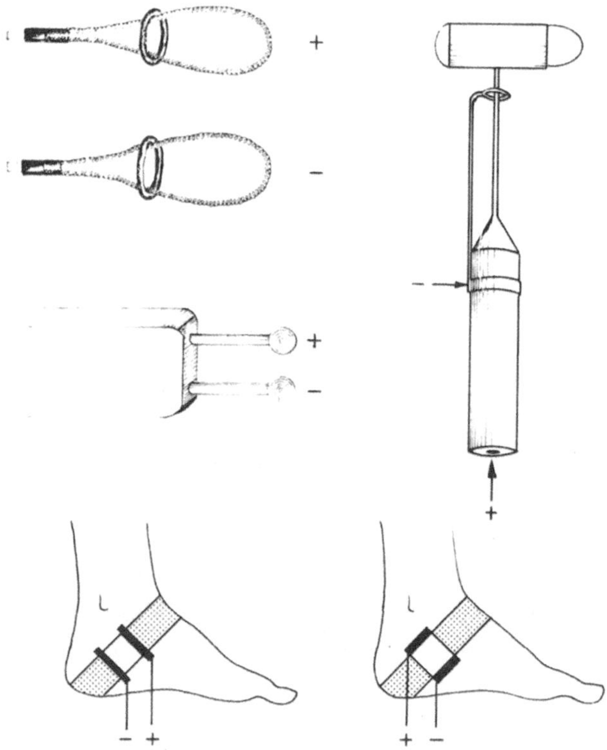

Fig. 5.6 Example of different electrodes used for somato-sensory (electrical or mechanical) stimulation.
For the mechanical stimulator the use of electrical conducting rubber may be preferable.

roots or in the spinal cord).
The second way of stimulation is direct stimulation of the mixed nerve (performed when one wants to know whether this nerve is abnormal).
With the third option one is especially interested in probable abnormalities of the sensory nerve itself, or in the processing of sensory information.
In this case only sensory nerves must be stimulated.
For the positioning of the stimulus electrode one may consult figure 5.1, in which the different sensory segments of the body and the different sensory nerve localisations are represented.
The most frequently used localisations are:

1. Median nerve on the index finger
2. Median nerve at the wrist
3. Peroneal nerve at the level of the head of the fibula
4. Sural nerve at the ankle

The numbers 1 and 4 are exclusively sensory nerves and 2 and 3 are mixed sensory and motor nerves.

3.4 Problems of neck and head muscles, eye artifacts and alpha rhythm

The abilities of an evoked potential department are well reflected in the quality of the EP's produced. This quality is influenced, among other things, by the signal-to-noise ratio. Noise in SSEP examinations consists especially of muscle artifacts, and movement artifacts. For near field and far field SSEP in the neck and head region, the muscle artifacts from the head and neck may be disastrous. We can avoid, or at least diminish, it by adequate, friendly patient care.
The attitude of the technician and the doctor are the highest promotors for rest or tranquillity. When the patient lies on a coach or sits in a very comfortable chair the artifacts neck muscles may also be diminished.
Muscle artifacts of the temporal muscles may be influenced by opening of the mouth.
Eye artifacts, as well as alpha rhythm, can be diminished by

asking the patient to open his eyes and fixate, for example, on a painting which is not too close by.
Stimulation of peripheral nerves may provoke somato-motor responses over the scalp. These myogenic responses may interfere with the neurogenic SSEP. Myogenic potentials with latencies of 15 ms and 100 ms are described in evoked potentials derived from the scalp. In general it is possible to minimize these potentials by relaxation. Take notice that these potentials may probably interfere especially in neurological patients (children).

3.5 Stimulus artifacts

These artifacts are sometimes highly disturbing because of their amplitude and duration. The only way to reduce these artifacts is by proper grounding and a careful positioning of both the stimulating and recording electrodes.

3.6 Problems with noise and light

What is described in 3.4 also holds for too much ambient noise or light. These may provoke muscle tensions in the patient. Rhythmical noise may provoke its own cortical reactions and may intervene with the SSEP. Light may not flicker for the same reasons. Too little noise and too little light, however, may give changes in the alertness of the patient, which may influence the amplitudes of the late components.

3.7 The SSEP-room

In general, the demands for the SSEP room are the same as for the departments of clinical neurophysiology, where electroencephalography is done.
It must be possible to enter the room with a bed. The light must be such that it can be dimmed (no fluorescent lighting), while the curtains of the room must be of such material that the room can be made dark. In the room there should be a comfortable settee. An alarm signal must be available in case of emergency.

Because of electrical and magnetical influence it is advisible to locate the room at a distance of more than 25 meters from radiological, surgical and physiotherapeutic departments. Also elevators and some clinical laboratories may have disturbing influences. A separate electrical main supply with properly grounded system may be necessary.

3.8 The optimal configuration

The most optimal position for a patient to undergo a SSEP is, in general, when he is lying on a relatively soft mattres. On the ceiling a point (for example a picture) is indicated where the patient can fix his eyes on, in a more or less comfortable way. This fixation diminishes the eye-movement artefacts. The lights on the ceiling may be very tiresome when they shine in the eyes.
A technician must always be in the neighbourhood. Repetitive stimuli (muscle for example) must be avoided. The temperature of the room should be above 21°C (or 69,8°F), and if possible a direct heating system should be directed to the place of stimulation.

3.9 Mental state

The amplitude of the late components of the EP is dependent on the mental state of the patient. All normal values are referred to a state of alertness, generally with eyes open. Some medications may influence the mental state of the patient (especially psychotropic drugs).

4. METHOD OF ANALYSIS

4.1 Equipment

For somato-sensory evoked potential examination the equipment must exist of a stimulating unit, some amplifiers, an averager and a display unit.
For detailed information see the second chapter of this book.

4.2 Sweep duration and sample frequency

It is useful always to analyse a prestimulus interval. A prestimulus interval over a period of 25% of the poststimulus interval is adequate. For short latency SSEP (until 45 ms) a sweep duration up to 100 ms may be chosen, for the late components a sweep duration of at least 400 ms is necessary. Sometimes these late components can be delayed up to 700 or 800 ms. The number of points on the sweep where the averaged signal is measured can best be chosen in such a way that the time difference between the two points is less than 4% of the measured time, which means that the measuring error will be less than 8%.

4.3 Peak detection

The latencies of the SSEP are measured as peak latencies. In general these peaks are indicated by means of visual inspection, which in itself gives, of course, an error. Automatical methods, as for example the detection of the point where the first derivative of the EP waveform crosses the baseline (after smoothing with an appropriate window) are generally not implemented in the various hardware equipment that is commercially available.

5. LOCALIZATION OF THE VARIOUS COMPONENTS

The SSEP may be derived from different parts of the body. The so-called distal sensory potential is found near the place of stimulation. The proximal and spinal sensory potentials are found near or on the vertebral column. The cortical SSEP is found at the skull, and is divided in specific and aspecific components. In figure 5.7 the form of the cortical SSEP is draw, in its most simple appearance. Many other components are described by different authors. The shape of this cortical SSEP (fig.5.7) is accepted by all of them.
The N_1 / P_1 of this cortical SSEP (until 45 ms) is the specific part and N_2 - P_3 (45 until 300 ms) is the aspecific part.

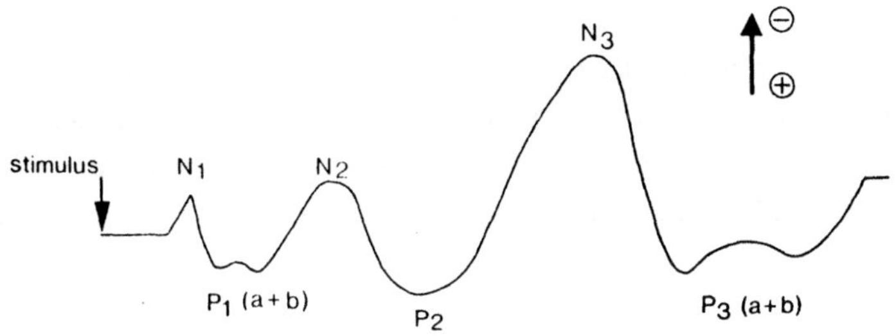

Fig. 5.7 The cortical SSEP (most simple form). N_1 and P_1 are "specific" and localized at the cortical recepting area, N_2 - P_3 are aspecific and found diffusely over the cortex.

5.1 Distal sensory

a. at the wrist

Stimulation of the digital nerves with ringelectrodes (cathode proximally) causes a SNAP at the wrist, for the median nerve if the index finger is stimulated and for the ulnar nerve if the little finger is stimulated.

One can use subcutaneous needle electrodes placed near the median or ulnar nerve, or surface electrodes. These different techniques yield different normative values.

b. at the elbow

When measuring the sensory nerve action potentials of the median nerve, the electrodes are to be placed on the skin in the cubital fossa.

The recording of the action potentials of the ulnar nerve can be done just below or above the olecranon in the ulnar sulcus.

c. at the ankle

Stimulation of the sural nerve behind the lateral malleolus will evoke a sensory nerve action potential.

5.2 Proximal sensory (see fig. 5.8).

At Erb's point in the supraclavicular fossa one can record a compound action potential which originates in the brachial plexus. The potential has a peak latency for the negative peak of 9 ms after stimulation of the median nerve at the wrist. The indifferent electrode can be located on the mastoid bone behind the ear or at FP_z.

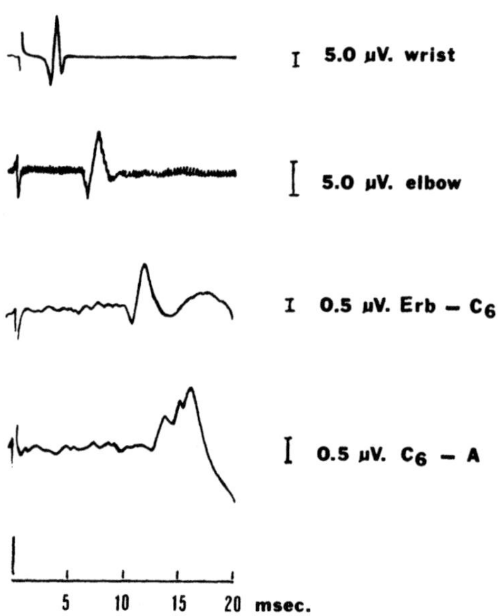

Fig. 5.8 Schematic drawing of sensory nerve action potentials following stimulation of finger III, the action potential at Erb's point and the cervical sensory evoked potential.

5.3 Spinal sensory (see fig.5.8 and 5.9).

a) At the cervical level electrodes can be placed on the spinal processes of the 7th till the 2nd cervical vertebra. The indifferent electrode is usually located at FP_z.
After stimulation of the median nerve at the wrist a negative potential is recorded with three negative peaks, the greatest of them with a peak latency of 14 ms. Chiappa (1980) mentions two peaks, an A wave and a B wave. The site of origin of these very small potentials is probably the dorsal ganglion in the dorsal root and/or the dorsal columns or the nuclei of these columns at the upper end of the cord.

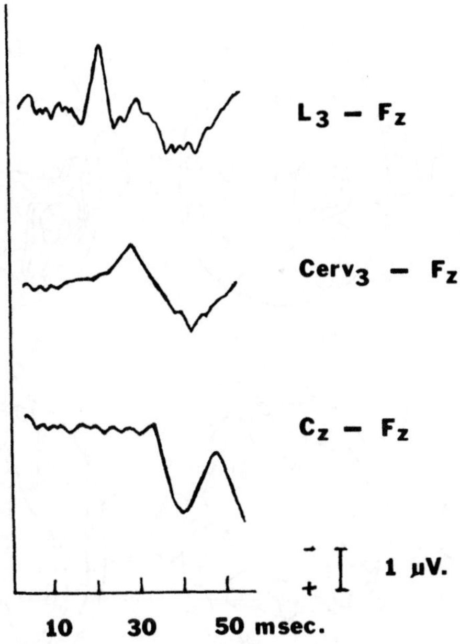

Fig. 5.9 Schematic drawing of two spinal evoked potentials and one cortical evoked potential following stimulation of the tibial nerve in the popliteal fossa.

b) After stimulation of the tibial or peroneal nerve in the popliteal fossa, the volley of action potentials will enter

the spinal medulla at the level of the first lumbal or twelfth dorsal vertebra (table 5.2).
A spinal evoked potential can be measured with an indifferent electrode at a higher level, for instance at the sixth dorsal vertebra.

5.4 Specific components (see fig. 5.10A and 5.10B).

Given the discrepancy between different authors about the number of specific and non-specific components of the cortical SSEP, we will give here a description of those components on which a relative agreement exists.

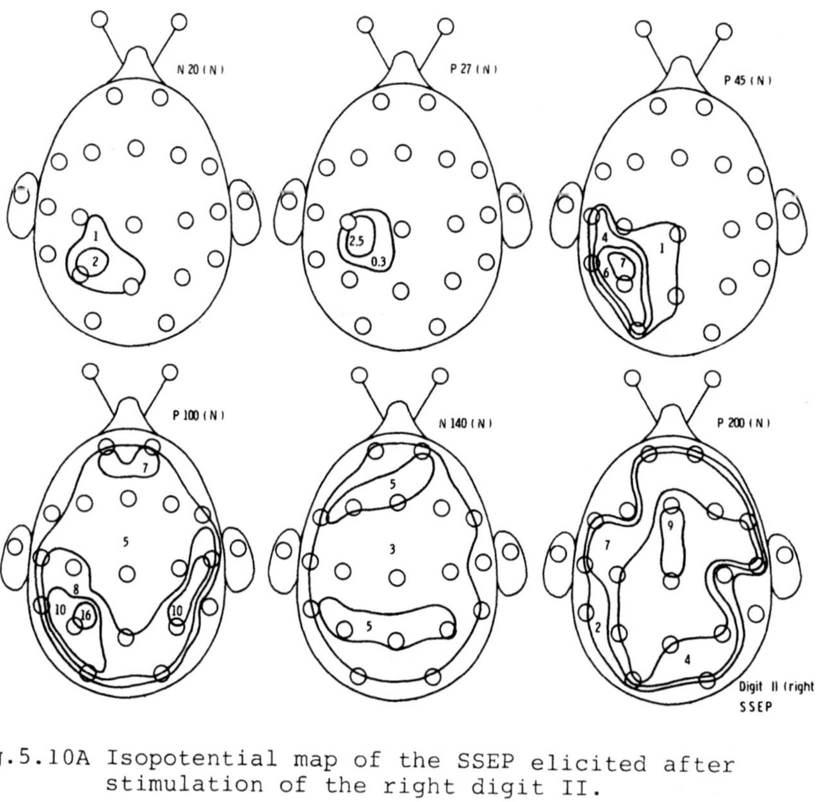

Fig.5.10A Isopotential map of the SSEP elicited after stimulation of the right digit II.

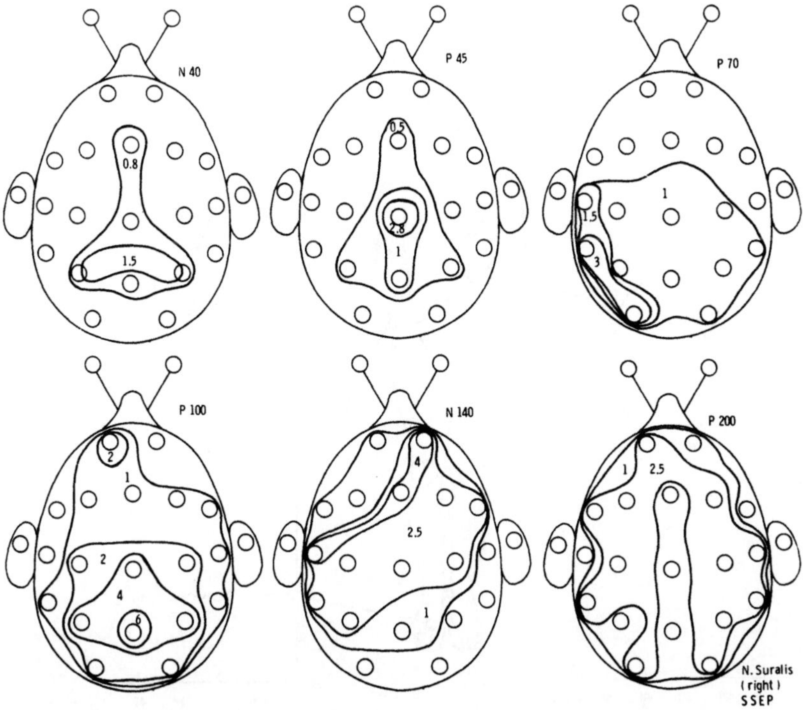

Fig.5.10B Isopotential map of the SSEP elicted after stimulation of the right sural nerve.

The place where the specific cortical reaction is maximal is found just above the sensory representation of the stimulated part of the body on the cortex (see figure 5.3). After stimulation of the sural nerve or peroneal nerve the maximum of the specific cortical reaction is found just 1 cm behind Cz. For the localization of electrodes on the head one is referred to the classical localization in Electroencephalography (10-20 system).

The maximum of the specific reaction after stimulation of the Median nerve at the index finger or at the wrist is found in between C3 and P3 or C4 and P4. This point is generally located 7 cm laterally from the midline, and 1 cm behind Cz. In figure 5.7 the localization of the specific components (N20, P27 and P45) is given after stimulation of the index finger.

5.5 Non-specific components of the SSEP (see figure 5.11).

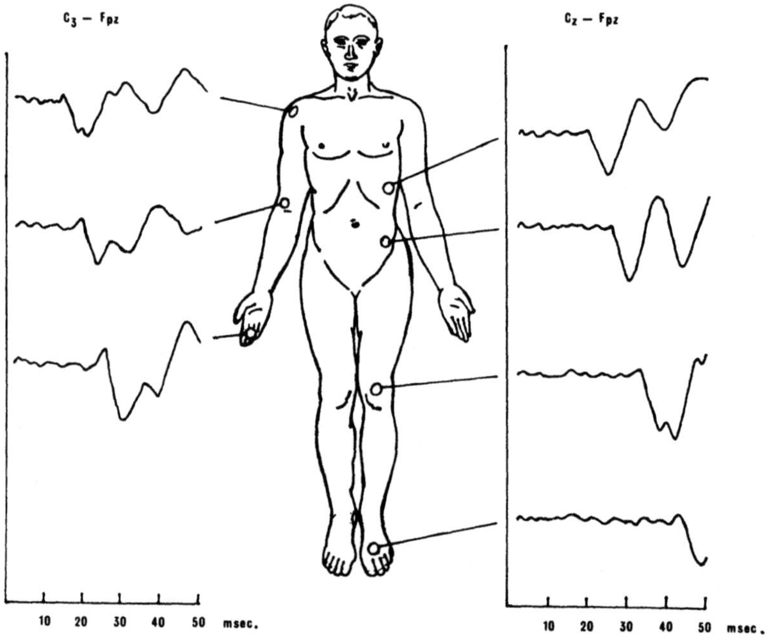

Fig.5.11 Examples of mechanically evoked SSEP's derived following stimulation of different parts of the body.

The late components are those components which are found in time after the generation of the specific cortical evoked potential. This specific reaction is only found in a very restricted area. The late components are found diffusely over the scull. In figure 5.11 the localization of the late components (P100, N140 and P200) is given after stimulation of the right index finger.

5.6 Generators and localization (a summary)

In this table a summary is given of the localization, nomenclature and probable generators of the different components of the cervical and cephalic SSEP (table 5.3, reference electrode on the ear lobe).

Localization (maximum)	Name	Probable generator
Cervical	N9	N. Medianus in axilla
Cervical	N11	Dorsal column or spinal cord entry
Cervical	N13	Brainstem or lemniscus
Cervical	N14	Brainstem or lemniscus
All cephalic electrodes	P15	Brainstem
Centro parietal	N18	Thalamo-cortical pathway
Centro parietal	N20	Somesthetic cortex
Centro parietal	P27	Somesthetic cortex
Precentral	P22/N30	?
Fronto-parietal	N35	Somesthetic cortex
Centro parietal	P45	Somesthetic cortex
?	N55	?
Frontal	P65/N70	Myogenic
All homolateral electrodes	P80	?
?	N90	?
All heterolateral electrodes	P100	Frontal myogenic,other neurogenic
All cephalic electrodes	N140	Thalamo cortical and commissura
All cephalic electrodes	P200	Thalamo cortical and commissura
All cephalic electrodes	N260/P300	?

Table 5.3.

5.7 Chronotopographical mapping

In the former part of this chapter (Fig 5.9) a chronotopographical map of components of the somato sensory evoked potential is given. These maps may be constructed in two ways. Most maps are generated as iso-potential maps over a number of electrodes at different times after the stimulus. Maxima and minima may be coloured, which gives a magnificent picture about the spread of evoked potentials over the hemispheres after somato sensory stimulation.

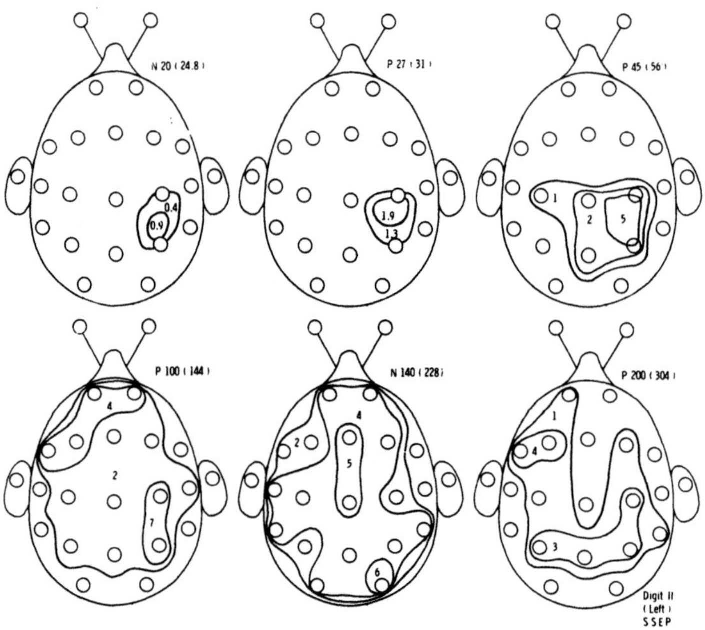

Fig. 5.12 Chronotopogram of the SSEP after left index finger stimulation in a patient with a CVA in the left hemisphere. Although the latency of the different components is abnormal, the chronotopogram shows in general a normal pattern, except for the N140 and P200. It is interesting to note that all late components are represented in the "healthy" as well as in the "CVA"-hemisphere.

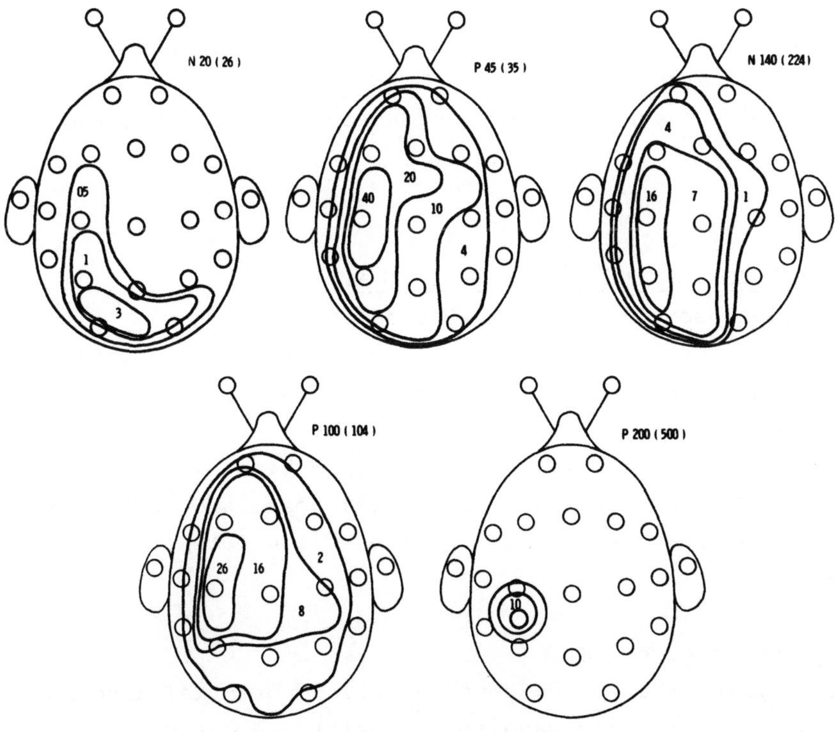

Fig. 5.13 Chronotopogram of SSEP after right index finger stimulation in a child with a ceroid lipofuchsinosis.
The distribution of the different components is highly abnormal, while the latencies are too large. This chronotopogram gives a clear view in the highly localized reaction of this pathologic SSEP.

Various examples of this method are given by Duffy et al. (1979, 1980 and 1981).
Another way to construct these chronotopographical maps is based on the assumption that the SSEP has a standard form over the cortex. One can identify the different components (for example the P100), measure this potential on the different electrode positions and in this way construct an iso-potential

mapping of these components.Thus one gets for example a P100 iso-potential map.The difference with the former method lies in the fact that P100 reaches the different electrodes (sometimes) at different times.
We will give two examples of a chronotopogram, constructed in this latter manner (Figure 5.12 and 5.13).

6. DESCRIPTION OF THE NORMAL SSEP IN MAN

6.1 Distal sensory (see figure 5.8 and 5.9).

The distal sensory nerve action potentials, after stimulation of the fingers, can be measured at the wrist, the elbow and even in the axilla.
The action potentials at the wrist have a biphasic or triphasic wave form. The amplitude in the cubital fossa and in the axilla decreases whilst the duration of the action potential increases.

6.2 Erb (see figure 5.8).

At Erb's point the evoked sensory action potential also has a biphasic or triphasic wave form, positive-negative or positive-negative-positive, with a negative main complex.

6.3 Spinal EP's (see figure 5.8 and 5.9).

The spinal SSEP in the cervical region, after stimulation of the median nerve at the wrist or the fingers, is mainly a negative wave with three distinct little peaks. The spinal SSEP in the thoraco-lumbar region, after stimulation of peripheral nerves at the lower limb, is most pronounced at the T12-L1 level. In the bipolar recordings the latency of the SSEP increases and the amplitude decreases if one measures the potential in rostral direction. Between L3-L1 and the T10-T8 electrode montage there is a phasereversal.
The amplitude of the SSEP can be augmented when both the left and right peroneal nerve are stimulated simultaneously.

6.4 Cortical EP (see figure 5.7 and 5.11).

The cortical EP consists of two distinct parts. The first part is the so-called specific cortical reaction, the second one is the aspecific cortical reaction.

The specific reaction is the part that is generated by the Thalamo-cortical tract and cortical somato sensory column reaction.

It consists of a negativety followed by one or more significantly positive parts. After finger stimulation, for example, a N20 and then a P27 and P45 may be found. This specific reaction is normally localized above the specific cortical somato sensory representation of the stimulated body area.

The non-specific reaction consists of at least a positive, negative-and-positive part which is found more or less diffuse over both hemispheres (see figure 5.11).

Many different peaks are described by various authors, but in general there is an agreement at least about the form as described in figure 5.11.

In human studies Goff et al. (1977) and Greenberg et al. (1977) both identify the components P100, N140 and P190, but they differ in their description of the components between P45 and P100. Goff and coworkers report 3 neurogenic peaks within this latency range, N55, P80 and N90, whereas Greenberg et al. only found a single negative peak N78.

Goff et al. (1980) derived median nerve wrist SSEP's from the pial surface and from probes inplanted in the depth of the brains of three patients. Before the P100 there is an indication of the existance of two small positive components (P65 and P80). The P65 and P100 (in frontal derivation) are composed of myogenic (but neural) potentials. The P80 and P100 are of somato-sensory neural origin. The P80 is probably larger on the homolateral, the P100 on the contralateral hemisphere. The depth-recorded vertex potential (N140/P190) is bilaterally symmetrical and a scalp-to-depth polarity reversal exists. The bilateral symmetry suggests a modality non-specific response arising from and/or mediated by a non-specific thalamo-cortical system or by commissural structures. Occasional discontinuities in depth recording of this activity suggest that we are dealing with more sets of neuronal processes which seem to yield the single biphasic latest somato-sensory evoked potential wave recorded at the scalp. The suggestion of these results of Goff et al. is that the origin of the latest components (N140/P200) are in some way related to the giant pyramidal cells with their long myelinated axons in the layers V and VI of the cerebral cortex. If this is true, then it must be possible by means of SSEP studies to differentiate between abnormalities in the upper cortical layers, which in the human species are relatively thick in relation to other species, and abnormalities in the lower cortical layers.

The latency of the first positive wave, after stimulation of the posterior tibial nerve at the level of the ankle, is about 6 ms shorter than after stimulation of the sural nerve at the same level.

Burke et al. (1981) conclude from their study that in man, group I muscle afferents from the lower limbs project to cerebral level and are responsible for this shorter latency of the cerebral evoked potential. Appropriate calculations suggest that most of the discrepancy in the latencies of the cerebral potentials evoked by posterior tibial and sural stimulation may be attributed to the differences in conduction velocity of the primary afferent fibres.

7. NORMATIVE VALUES

7.1 Distal sensory

The sensory nerve action potential at the wrist, after stimulation of the index finger with ring electrodes, has a latency of 3.5 $\pm$ 0.1 ms and an amplitude of 24 $\pm$ 3 microvolt (Buchthal and Rosenfalck, 1966).

In the cubital fossa the amplitude of the action potential is 6 $\pm$ 1 microvolt and the onset latency of about 8 ms.

Table 5.4

site of stimulation	Author	Erb's point		Cervical SEP		Cortical SEP	
		ms	µV	ms	µV	ms	µV
Median Nerve wrist	Chiappa	N8.4	2.7 ± 1.5	N11.8 Awave N13.7 Bwave	2.35± 0.9	N19.1± 0.8	3.5± 0.3
	Eisen	N10	5.0	N14	1	N20	1
finger III	Desmedt	N12	2.0	N12 N15 N18	1	N22	5

Peak latencies of the first negative peak of the EP at various sites of recording after stimulation of the upper extremity.

Table 5.5

Site of stimulation	Author	dorsolumbar level		cervical level		cortical SEP	
		ms	µV	ms	µV	ms	µV
popliteal fossa Tibial N.	Cracco	N11	2.0	N15	0.25	-	-
ankle Sural N.	Eisen	N21	4.0		-	N33 P40	-
	Jones	N22.6 ± 1.4	1.5 ± 0.6	N29	0.5	-	-

Peak latencies of the first negative peak of the SSEP after stimulation of the lower limb.

7.2 Proximal sensory (Table 5.4)

At Erb's point the latency of the action potential, after stimulation of the median nerve at the wrist, is ± 9 ms for the negative peak and has an amplitude of 2.7 ± 1.5 microvolt. Stimulation of the fingers will give a peak latency at Erb's point of 12 ms with a much smaller amplitude which is hardly measurable after exclusive stimulation of the little finger (Desmedt, 1981).

7.3 Spinal sensory (cervical)

The cervical SSEP has a peak latency for the main negative peak of about 14 ms and an amplitude of about 3 microvolt. According to Chiappa the interpeak latency from the action potential at Erb's point to the A-wave of the cervical SSEP is 2.1 +/- 0.3 ms and to the B-wave 4.1 ± 0.2 ms.
The peak latency of the cervical SEP recorded over the spinal processes of C7 is 0.95 ± 0.15 ms shorter than the same SEP measured at C2, which according to Desmedt is the conduction time in the cervical dorsal columns. Over a distance of about 55 mm this gives a conduction velocity of 58 m/s in the spinal cord.

7.3. At the ankle

Stimulating the sural nerve behind the lateral malleolus will evoke a sensory potential ± 12 cm above the ankle in the posterior midline of the leg of 5 - 20 uV, with a peak latency until 4 ms.

7.4 Spinal sensory (lumbal) (Table 5.5)

Measurement of the spinal SEP in the thoracolumbar region after stimulation of the tibial nerve in the popliteal fossa gives a latency of about 14 ms. Stimulation at the ankle will result in a latency of 21 ms for the same negative peak. The amplitude of this negative peak is about 2.5 microvolt.

7.5 <u>Age dependency of short latency SSEP</u>(Table 5.6)

Site of Stimulation

	Finger III					Tibial posterior nerve			
	Neck SEP		Cortical SEP			spinal EP		cortical EP	
age	onset	N	onset	N	P	onset	N	onset	P
first week	-	-	15	30	-	-	-	30	60
6 months	-	-	-	21	-	-	5	-	-
16 months	-	-	-	20	33	-	-	-	-
young adult	13	17	-	22	26	9	11	33	40
80 years	13	17	-	25	32	-	-	-	-

Latencies of the onset and negative and positive peaks of the spinal and cortical SSEP in relation to age.

7.6 <u>Cortical sensory</u>

<u>Normative values in newborns</u>:
Desmedt et al. (1979) described a N_{31} in newborns, which is equivalent to the N_{22} in adults after finger stimulation. From the plots in his publication it seems that the adult value of 22 msec of the first negative cortical potential is already reached within one year of age.

Normative data by Lüders, (1970) (Table 5.7)

-Sixteen males between 46-70 years old

-Median nerve, disc electrodes

-0.1 ms pulses of 10V above motor threshold

-1.25 sec interval

-C_3, reference on contralateral ear lobe

-TC = 0.1, HF not given

-1000 responses averaged

	mean latency (SD)*	mean amplitude (SD)*
N_1	17 (0.2)	1.4 (0.05)
P_1	24 (0.6)	1.4 (0.06)
N_2	30 (0.4)	1.5 (0.04)
P_2	44 (0.8)	1.6 (0.05)
N_3	62 (2.5)	1.3 (0.06)
P_3	102 (4.7)	1.3 (0.05)
N_4	142 (6.0)	19.3 (1.72)
P_4	229 (9.8)	21.8 (2.87)
N_5	384 (8.1)	-

*In groups of 19-29 years old the results are not significantly different.

Normative data by Tamura et al. (1972) (Table 5.8)

-Twenty males (9-18years old)

(for method see Normative data by Lüders (Table 5.7)

	mean latency (SD)	mean voltage (SD)
N_1	18 (0.28)	1.6 (0.05)
P_1	25 (0.68)	1.3 (0.07)
N_2	32 (0.99)	1.2 (0.07)
P_2	42 (0.84)	1.5 (0.04)
N_3	56 (1.39)	1.2 (0.08)
P_3	85 (2.3)	1.6 (0.07)
N_4	136 (4.0)	48.6 (4.52)
P_4	242 (8.8)	46.2 (4.62)
N_5	392 (13.5)	

Normative data by Tsumoto et al. (1972) (Table 5.9)

- 33 male and 8 females between 19 and 34 years old
- Disc electrodes to left lateral popliteal nerve near the caput fibulae
- 0.1 ms, rectangular pulses above sensation threshold
- Random stimulation with intervals between 1.5 and 3 sec
- Measurement made 2 cm behind Cz, reference left ear lobe
- Filter setting between 1 and 1000 Hz
- 100 responses averaged

	Mean (peak latency)	SD
P_1	34	2.0
N_1	45	3.3
P_2	56	4.7
N_2	74	5.6
P_3	99	7.3
N_3	126	15.0
P_4	204	16.0

Voltage (in μV)

P_1/N_1	4.0	1.1
P_2/N_2	4.5	2.3
P_3/N_3	5.1	2.8
N_3/P_4	11.0	4.5

<u>Normative values Nakanishi et al. (1974)</u> (Table 5.10)

- 9 male and 24 females between 19 and 32 years old
- pin-prick with an insect pin and touch stimulation with a soft brush (2 ms, rectangular vibration, latency from the beginning to the peak is 3.5 ms) and electrical stimulation (parameters not given)
- <u>Phalanx of middle finger or big toe</u>
- for electrical stimulation the median nerve and sural nerve were stimulated.
- electrodes on C_3 or C_4 and C_2, respectively
- filters are not given
- 200 responses averaged

Finger/median nerve	Electrical (SD)	Touch (SD)	Prick (SD)
N_1	18 (0.8)	29 (3.3)	27 (2.5)
P_1	25 (2.5)	36 (4.5)	36 (4.0)
N_2	34 (3.8)	45 (4.1)	44 (5.1)
P_2	44 (5.2)	56 (4.4)	56 (4.1)
Big toe/sural nerve			
N_1	33 (2.6)	54 (8.3)	43 (7.3)
P_1	42 (3.9)	60 (8.1)	50 (7.5)
N_2	53 (5.7)	72 (9.0)	60 (9.3)
P_2	65 (5.2)	80 (8.1)	70 (10.9)

<u>Normative values by Eisen and Elleker (1980)</u> (Table 5.11)

- 20 - 34 subjects in between 17 and 72 years old
- electrical stimulation (0.2 ms, 5/s)
- C_3 and C_4 and C_z, reference on FP_z
- filters 0.5 - 200 hz
- at least 1000 stimuli

	P_1	N_1	P_2
Median nerve	19.4 ± 1.3	22.5 ± 1.1	29.4 ± 1.5
Ulnar nerve	19.3 ± 1.5	22.0 ± 1.4	28.7 ± 1.5
Radial nerve	16.4 ± 1.3	19.2 ± 1.4	28.2 ± 1.6
Musculocutaneal nerve	14.1 ± 1.3	17.4 ± 1.2	25.4 ± 1.4
Sural nerve	-	35.1 ± 1.9	42.1 ± 1.4
Peroneal nerve	-	33.1 ± 2.2	39.9 ± 1.8
Saphenous nerve	-	36.8 ± 2.8	43.4 ± 2.2

<u>Normative values by Jones (1981)</u> (Table 5.12)

- 14 volunteers between 18-33 years old
- electrical stimulation (0.2 ms, 10-15 mA, 2/s regular)
- <u>Median nerve (wrist)</u>
- filters 0.2 - 5000 Hz
- at least 400 stimuli

		Latency	Amplitude
Erb	N9	9.4 ± 0.8	8.2 ± 4.6
Cervical	N13	12.7 ± 0.9	2.6 ± 1.0
Cortical	N17	17.2 ± 1.1	0.7 ± 0.3
	N18	18.3 ± 1.2	1.3 ± 0.6
	P21	21.3 ± 1.3	5.2 ± 3.9
	P32	31.5 ± 4.7	1.6 ± 1.1
	P45	44.5 ± 4.4	3.1 ± 2.1
	N55	57.1 ± 8.2	1.7 ± 1.4
	P80	79.9 ± 11.0	2.3 ± 1.3
	N130	132.0 ± 21	6.1 ± 2.0
	P190	187.0 ± 21	2.0 ± 0.6
	P300	298.0 ± 19	2.8 ± 1.8

Normative values by Rossini et al. (1981) (Table 5.13)

- 13 females and 9 males, mean age 24 years old
- electrical stimulation (0.2 ms, 9-10/s)
- Peroneal nerve (popliteal fossa)
- electrode C_z, reference on FP_z or A_z
- filters 5 - 3000 Hz
- 2000 stimuli

		Peak latency	Voltage
$P1_A$	(only ear reference)	17.0 ± 2.0	0.04 - 0.2
$P1_B$	(only ear reference)	19.2 ± 1.6	0.04 - 0.2
$P2_A$		21.5 ± 1.8	0.04 - 0.02
$P2_B$		23.4 ± 1.7	0.1 - 0.4
$P3_A$		25.6 ± 1.8	0.1 - 0.4
$P3_B$		27.7 ± 1.7	0.04 - 0.2
N1		28.7 ± 2.0	0.25 - 1.0

Normative values by Riffel and Stör (1982)

- 18 females and 12 males, mean age 34 years old
- electrical stimulation (0.1 ms, 5/s)
- Tibial nerve (ankle)
- electrode C_z^{1}, L_1, L_5 and Cerv. 2, reference on F_z or pelvis
- filters 10-1000 Hz cortical, 50-1000 Hz spinal derivation
- 2048 stimuli

	Peak latency	Voltage
L_5	18.0 ± 1.2	0.3 ± 0.2
L_1	21.2 ± 1.6	0.6 ± 0.3
Cerv. 2	28.8 ± 1.9	0.6 ± 0.3
C_z^{1} positive	37.9 ± 2.0	1.8 ± 1.3

Normative values (own material)

- 10 volunteers between 20 and 30 years
- electrical stimulation (0.1 ms, 10-15 mA, regular 3/ s for specific components, random 2-4 s for non-specific components)
- Median nerve (wrist), index finger, sural nerve (ankle)
- filters 1-500 Hz
- for specific 1000 stimuli (twice)
  for non-specific 100 stimuli (twice)

Normative values (mean and SD)

	right index finger		right wrist	
	latency	voltage	latency	voltage
P18	18.1(1.6)	1.7(1.2)	15.4(1.0)	1.7(1.6)
N20	22.4(0.8)	2.3(1.3)	19.7(0.9)	2.4(1.5)
P27	24.4(9.0)	1.0(0.9)	25.3(3.0)	1.0(0.9)
P45	40.0(6.1)	1.0(0.7)	39.7(5.7)	0.8(0.5)
N60	59.7(6.9)	3.0(2.8)	61.6(5.1)	2.4(2.2)
P100	104(11)	3.3(2.4)	97 (14)	4.4(2.9)
N140	140(18)	3.8(2.7)	138 (15)	6.1(4.4)
P200	177(21)	5.2(3.3)	177 (17)	3.8(4.7)
P300	274(11)	2.8(1.8)	279 (28)	3.1(1.7)

Right Sural Nerve (mean and SD)

	latency	voltage
N35	38.2(2.4)	-
P40	45.4(2.7)	-
P60	71.7(6.0)	-

Cervical-thalamus (N.med. right) : 1.7 ± 0.6
Thalamus-cortical (N.med. right) : 4.4 ± 0.37

(this "central conduction time" remains constant until 49 years and increase with 0.3 ms between the fifth and sixth decade, whereafter it remains stable (Hume et al. 1982)

8. CLINICAL USE OF THE SSEP

In clinical practice the SSEP is needed for two distinct reasons. The first reason is the examination of sensory propagation along the myelinated pathway from the place of stimulation to the cerebral cortex. This means that the SSEP is useful in diseases where lesions may be found in the nerve, the plexus, the cervical roots, the cervical column, or the brainstem. Somato-sensory EP's may be derived from different points of this pathway and if necessary compared with EP's obtained after stimulation of the heterolateral side of the body. With this method one is able to locate the lesion in this pathway and to get an impression about the severity of this lesion.

The second reason is the examination of the propagation of sensory information about the cortical hemispheres. These are reasons to expect that these late components are changed at least when the cortex itself is affected (Colon et al. 1979, Goff et al. 1980, Kazis et al. 1982).

Disturbances in the communissial fibres may give propagation defects of the late components to the homolateral hemisphere (Ferris 1973, Goff et al. 1980, Kazis et al. 1982)

So diseases that are especially bound to the cortical and sub-cortical neurons ("gray matter") or the large interconnecting cerebral fibres cause changes in the late components of the SSEP. Especially in diseases where diffuse dysfunction of the cortical gray exists or begins, the late component changes seem more sensitive than the EEG. The advantage of the EP over the EEG is also found in the fact that the EP is relatively simple in its form and distribution, whereas it is rather easy to quantify this potential and compare it with normative values.

Today little is known about the ability of the SSEP to detect abnormalities in intra- or inter-hemispheric interaction.

In the near future this may well be a new and intriguing field of examination and diagnosis.

8.1 Distal sensory

In many forms of polyneuropathy there are complaints about sensory disturbances in hands and feet and there are clear abnormalities of the sensory system during the neurological examination.

The motor conduction velocity and/or the sensory conduction velocity can be diminished. The distal latency can be prolonged. Accordingly the latency of the SSEP, is increased. The form of the SSEP can be abnormal and the amplitude very small, or even totally absent.

In all these cases of polyneuropathy the peripheral sensory nerve action potential and the measurement of the sensory conduction velocity will give more relevant information than the cortical SSEP can do.

There are several local causes for neuropathy, especially pressure or entrapment. For instance, the median nerve can be compressed in the carpal tunnel with increased conduction time of both the motor and sensory distal latency.

In traumatic lesions of the peripheral nerves, the recording of the SSEP may give some information about the severity of the nerve lesion, whether total or subtotal, and the process of healing can be followed and objectivated in time.

8.2 Proximal sensory

Lesions of the dorsal roots of the spinal nerve, as in root-avulsion, can be examined by the SSEP recordings.

Anzinska and Cracco (1980) give some examples of these applications.

If the nerve lesion is distal to the dorsal ganglion in the dorsal root, then the peripheral axon will degenerate; in this case both the peripheral sensory conduction velocity and the SSEP will be abnormal or even absent. If on the other hand the nerve lesion is located more proximally between the dorsal ganglion and the spinal cord, then the SSEP is abnormal but the peripheral sensory conduction will remain normal.

8.3 The spinal SSEP

The most severe lesion of the spinal cord is the total transsection. All sensory conduction is blocked at that level.
In subtotal lesions the SSEP can give some information about the preserved function of the dorsal columns.
In unilateral transsection of the spinal cord (Brown Sequard Syndrome) the SSEP will be normal at the side with normal gnostic qualities of the sensory system and absent on the other side of the body below the level of the lesion.
During operations the SSEP can give information on the spinal cord or vertebral column (Lesser et al. 1981).
The level of the spinal cord lesion can be examined also by means of the SSEP according to the dermatomes that show a sensory disturbance (see fig. 5.1), (Stöhr et al. 1982).

8.4 Specific cortical EP

- Demyelinating disease

A huge amount of literature can be found on changes of the form and latency of cortical-derived SSEP's. In all studies a latency shift of the N20 is mentioned, a shift which is correlated with abnormalities in the myelinated pathway in the central nervous system.
Different authors can be mentioned: Halliday, 1963, Baker, 1968; Namerow, 1968; Colon, 1977; Matthews, 1979A and B; Eisen, 1980; Desmedt, 1979; Trojaborg, 1979; de Weerd A., 1981 and Yamada, 1982, Anziska, 1980 and many others.

In 1979 Matthews and Small tried to relate the changes in the clinical state of patients with M.S. (sclerosis multiplex) to abnormalities in the SSEP. The short latency somato-sensory evoked potentials, whether normal or abnormal, remained stable in patients without relapse. The potentials became more and more abnormal in the majority of those with a relapse involving the spinal cord. The latency shift of the N20, therefore, can give an indication about the damage of the long fibres in the spinal cord in this disease.

The localization of the lesions in the central nervous system may be detected by measuring SSEP in different regions of the lumbar trunc and cortex. Colon et al. (1977) gave an example of this ability of the EP of the somato sensory system. The results of this investigation indicated that localization of central nervous system pathology is possible. The method is useful in addition to the tools that are used to evaluate quantitatively the results of different treatments in this type of disease.
Changes in the latest components, which might indicate abnormalities in the cortical system, have not been studied systematically.

Bruggenhout et al. (1982) analysed the value of SSEP, VEP and BAEP for the detection of clinical and subclinical CNS lesions in 100 patients suffering from multiple sclerosis. He concluded that all clinically manifest posterior column lesions are associated with abnormal SSEP.
However, a substantial proportion of clinically evident lesions in the visual pathway or the midbrain and pons are not detected by the VEP and BAEP.

- Friedreich's ataxia

Patients with Friedreich's disease show changes in their peripheral nerve fibre distribution.
The largest fibres (18-12 μ in diameter) are not present. In EMG generally no or only low sensory distal potentials are found.
Colon et al. (1978) have been able to demonstrate the ability to elict a cortical somato-sensory evoked potential in patients with selective A-alpha sensory fibre disease. The remaining A and C fibres appear to be sufficient to evoke a potential in the cortex, but the propagation velocity is diminished.

- Spino-Cerebellar Degeneration

In late Spino-Cerebellar Degeneration the features of the cortical SSEP after finger stimulation are not different from those found in Friedreich's patients. With moderate stimulation a rather late SSEP is observed (Noël et al. 1980).This might be the result of the loss of neurons that is sometimes observed in the N.Gracilis and Cuneatus or atrophy of the medial lemniscus.

- Trauma/death

Scalp recorded, short latency somato-sensory evoked potentials (potentials from the brachial plexus, brainstem and the N20) after median nerve stimulation were elicited in "braindead" patients (Anziska et al.,1980). Potentials that arose in central subcortical afferent pathways were recorded in about 70% of the patients, which suggests that these pathways are capable of conducting impulses in some clinically braindead patients.
Goldie (1981) concluded that the SSEP has greater clinical utility than BAEP in braindead diagnosis because medullary components may be elicited which indicates that the input signal reaches the central nervous system. None of the braindead patients had subsequent (SSEP) waves.
These results of Anziska and Goldie must be interpreted with care, the number of patient-results is rather low.
In comatose patients following closed-head injuring Torre et al. (1978) demonstrated the usefulness of SSEP as a prognostic predictor. The majority of negative recovery patients only showed SEP specific components consisting of only two peaks. This study determined that the late components of the SEP are more critical than the early components, and that the return of consciousness and improvement seems to be correlated with the recovery of the late components.

- Vascular disturbances

Branston et al. (1976) measured the degree of recovery of the SSEP following a period of up to 65 minutes of ischemia of the middle cerebral artery in baboons. Failure of recovery was associated with significantly greater depth of ischemia and tissue hypoxia during occlusion, and with significantly greater and persisting tissue hypoxia after occlusion than complete recovery.

Reduced postocclusive PO_2 levels associated with incomplete recovery of the evoked potential suggest that reduced perfusing during ischemia was sufficiently severe to cause some degree of irreversible anoxic damage.

In patients with anoxic lesions in the thalamic region, the SSEP has a reduced amplitude and increased latency on the affected side (Noël et al. 1975)

In patients with a Wallenberg syndrome or Weber syndrome, no changes in the SSEP can be found, while in patients with a locked-in syndrome the evoked potentials present marked bilateral abnormalities. An example of the change in SSEP after CVA is given in figure 5.14.

- Hypomyelinosis and leucodystrophy

The SSEP is delayed (especially the N20 and P27/N45) in patients with diseases involving the myelinated pathways. The more severe the disease, the less an EP is found in the cortical derivations. Colon et al. (1979) and Markand et al. (1982) could demonstrate this in patients with leucoencephalitis and subacute necrotizing encephalomyelopathy. Also, a patient with Moya-Moya disease was described in which the SSEP was not elicitable.

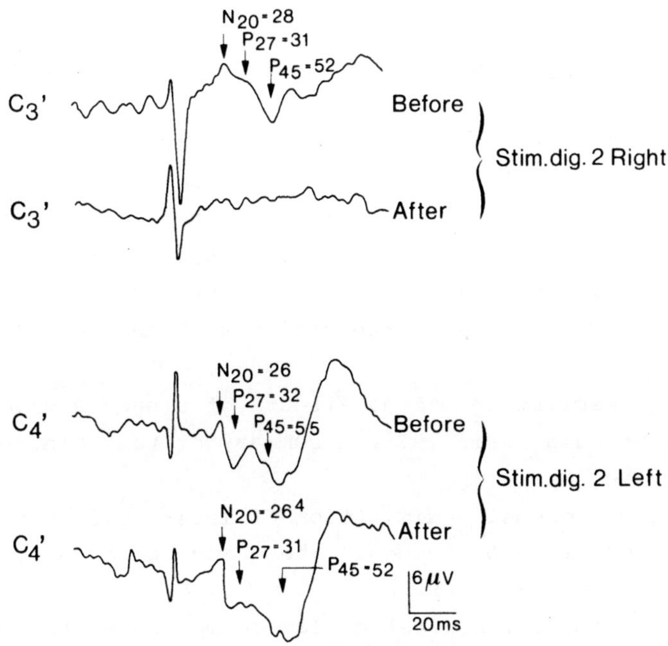

Fig. 5.14 Example of a SSEP after right and left index stimulation before and after CVA in the left hemisphere.
Before the CVA both N20 on the right and left side are too slow. After the CVA a N20 is not elicited in the left hemisphere.

- Vit B12 deficiencies

Abnormalities of somato-sensory evoked potentials were demonstrated by Krumholz et al. (1981) in two out of seven individuals with vitamin B12 deficiency. The evoked potential prolonged latencies correlated directly with the degree of neurological dysfunction. Abnormalities were present in sensory systems without clinical evidence of involvement and were similar to those found in individuals with multiple sclerosis.

8.5 Late components (sometimes combined with specific components)

- medication

The effect of xylene and alcohol consumption on SSEP is minor or absent (Seppä-äinen, 1981)
The early potential of the SSEP, which is not influenced by Kaimic Acid (destroys the cortical neurons), is similar to that of the field potential recorded by tungsten needle electrodes in VPL nucleus of the thalamus (Tamaki et al.1981).

An extensive description of the effects of drugs is given by Shagass and Straumanis in 1979. A summary of the relevant changes is given here:

Barbiturates: progressive diminution of later component amplitudes and no change in the early components.
Diethyl ether: see barbiturates.
Cyclopropane: diminution first of late components, then of early components.
Nitrous oxide: decrease of late components and early components, with high dosis.
Fluroxene: see nitrous oxide.
Methoxyflurane: see barbiturates.
Benzodiazepines: diminution of amplitude, chlordiasepoxide may increase the latency of the early components.
Amphetamine: this may shorten the latency of the various components.
Lithium: increase of the amplitude of the early components.

Tricyclic drugs: decrease of the latency of the early components and increase of the latency of the latest components.
Alcohol: reduction of amplitude of the late components.

From different studies it seems possible that patients with CNS diseases may react differently on these drugs.

- Effect of sleep on SSEP

As yet no extensive studies of the effect of sleep on the wave-form and latency of the SSEP have been made. Some studies indicate changes in the latency of the late components and changes in the distribution of the different components over the scull. These studies suggest that SSEP studies must be done under standard conditions, especially for vigilance.

- Gray matter diseases a.o. Lipofuchsinosis

In gray matter diseases the secundary complex of the SSEP, especially of the latest components, may be disturbed. This phenomenon is seen in patients with ceroid lipofuchsinosis, mucopolyacharidosis and SSPE/phase 1 (Colon et al. 1979).
Laget et al. (1976) analysed the SSEP in 43 children with hemiplegic cerebral palsy. In all children there was a positive correlation between the SSEP changes and the affected side of the body. In 12 children the SSEP in the damaged hemisphere disappeared.
In 31 children the SSEP was less developed in the damaged hemisphere than the potential of the "healthy" hemisphere.
It was concluded that SSEP gives more accurate and reliable diagnostic indications than the EEG does. SSEP findings allow cerebral damage to be suspected earlier than the EEG results and therefore are of greater prognostic value.

In Huntingtons chorea the latencies of the SSEP are slightly prolonged, less marked in the early peaks and more pronounced in the later ones (Oepen et el. 1981). In these patients the components after 70 ms are often reduced or even absent, and interhemispheric asymmetries may be found.

Dawson (1947b) first reported that abnormal cortical potentials could be evoked in the Rolandic area by stimulation of the peripheral nerves in the contralateral arm or leg in a patient with myoclonic epilepsia. He found that these potentials appeared to be part of an abnormal enhancement of much smaller potentials which could be evoked in healthy subjects, but which were only a fifth to a tenth of the size of those in myoclonic patients. The majority of patients with progressive myoclonic epilepsy show a high amplitude somato-sensory evoked potential in response to electrical stimulation of the median nerve. The N33 component of this high amplitude SSEP is found to be similar to the myoclonus related cortical spiked in its waveform, time relationship and topographical distribution (Shibasaki,1978) This suggests an involvement of similar physiological mechanisms in the genesis of both phenomena.

Tackman et al. (1979) described different neurological diseases. His results were as follows:

Metachromatic leucodystrophy	no SSEP on record
Globoid cell leucodystrophy	" " " "
Tay Sachs disease	" " " "
Mucopolysacharidosis	" " " "
Neuronal ceroid lipofuchsinosis	
Hogberg type	no SSEP on record
Bielschowsky type	markedly enlarged
Spielmeyer type	no SSEP on record

- Psychiatry/Psychology

In patients with hysterical anaesthesia the contralateral somato-sensory response to strong stimuli on the affected side differ from a habituation pattern in that they show fascilitation rather than habituation (Moldowsky et al. (1975). This seems more or less in contrast with the findings of Levy et al. (1973), who found a significantly smaller amplitude of EP's from the affected area, which they concluded, would indicate a centrally acting inhibitory mechanism. Shagass et al. (1980) reported differences in the topography of the different SSEP components in patients with psychotic depression and chronic schizophrenia in relation to normals.

The SSEP may be altered by human activity.

Rushton et al. (1982) could demonstrate that the N20-P30 primary cortical response is only moderately affected by task content, while the P45-N55 secundary response is markedly "gated" by movement of the stimulated digit.

8.6 Miscellaneous

- Maturation

Evoked potential study is useful in the analysis of cerebral maturation. Hrbeck (1977) demonstrated that the most characteristic features of evoked potentials in asphyxiated infants were abnormal response patterns and increase of latency. Repeated observations are important, a permanently abnormal EP is a serious prognostic sign.

9. EXAMPLES OF SSEP IN DAILY PRACTICE

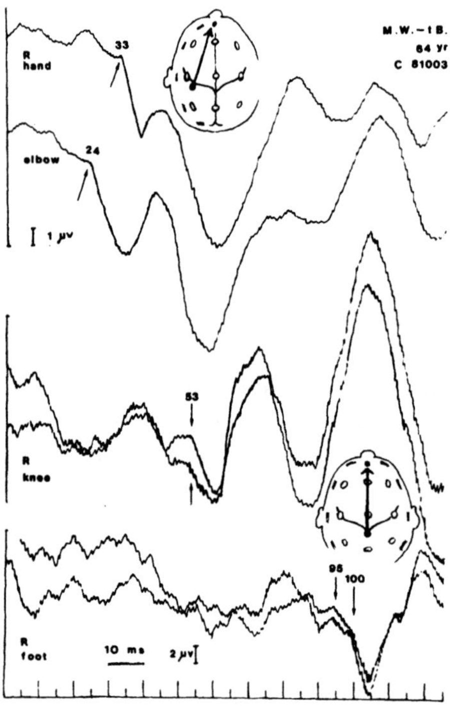

Fig. 5.15 Mechanically evoked response (tapping) from the right arm and the right leg in a 64-year-old woman with uremic polyneuropathy caused by renal failure.
The latency of the SEP from the right foot is particularly long. Each curve is the result of 200 stimuli.

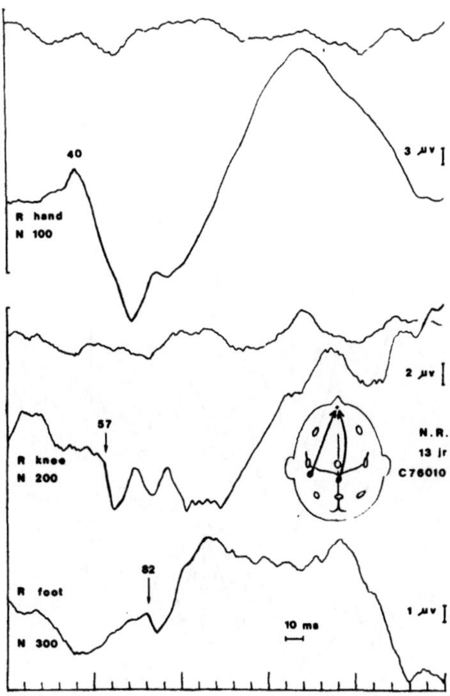

Fig. 5.16 13-year-old boy with hypertrophic polyneuropathy (type Déjérine Sottas). Mechanically evoked SEP from right hand, knee and foot.
Reference F_{pz}. Motor conduction velocity of ulnar nerve and peroneal nerve are also slowed down, 23 m/s and 36 m/s, respectively.

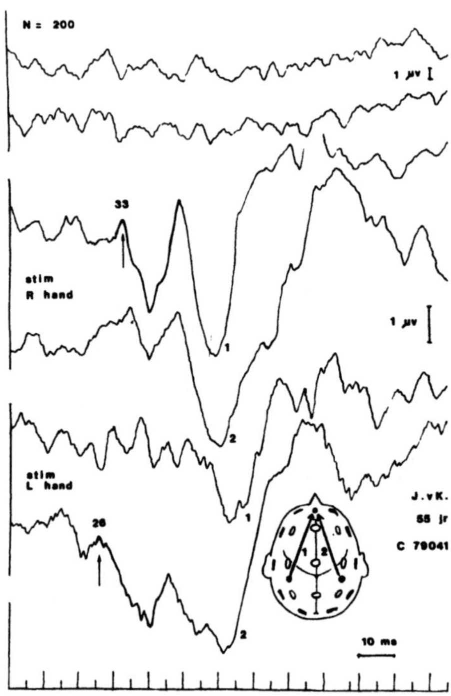

Fig. 5.17 55-year-old man with very small ischaemic lesion in the left thalamus. SSEP following mechanical stimulation of the right hand with increase of latency (33 ms) as compared to normal latency (26 ms) following stimulation of the left hand.

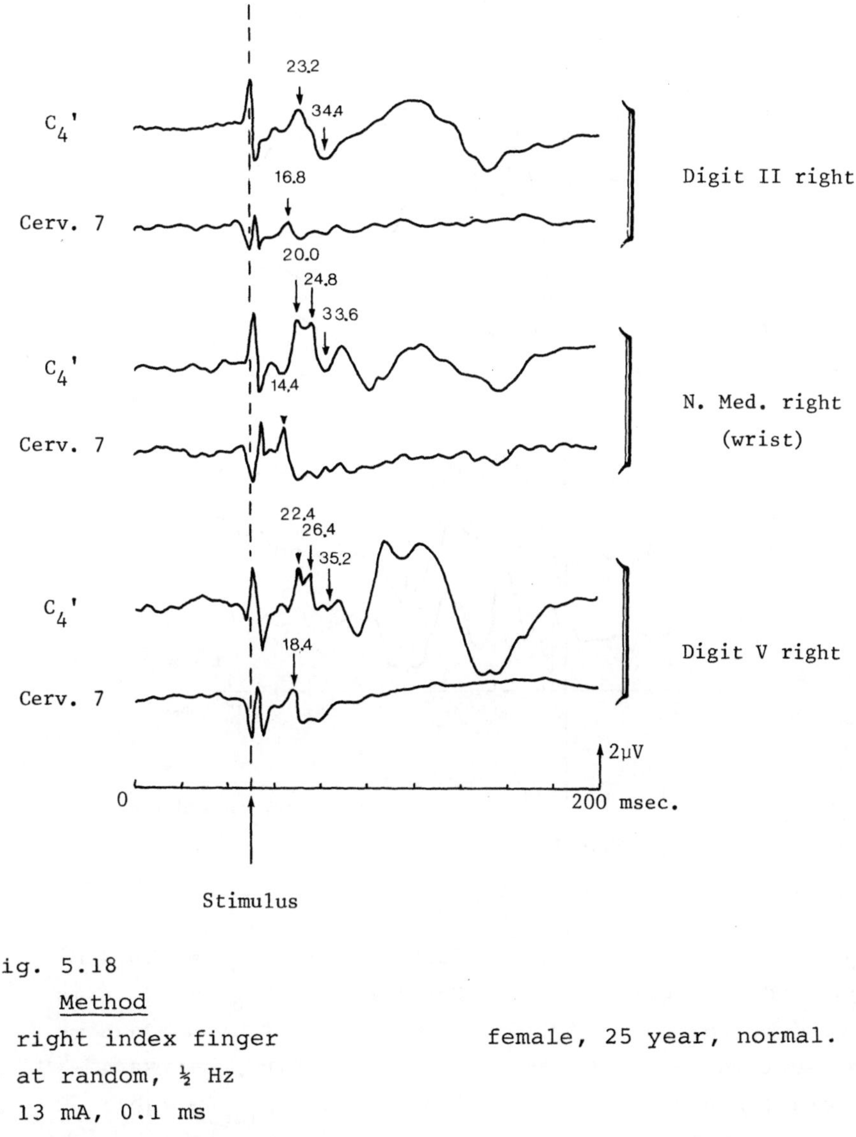

Fig. 5.18

Method

- right index finger
- at random, ½ Hz
- 13 mA, 0.1 ms
- filters 1 - 250 Hz
- reference A2

female, 25 year, normal.

Conclusion: example of the shape of the SSEP in neck and cortical region after stimulation at different localizations of the right hand.

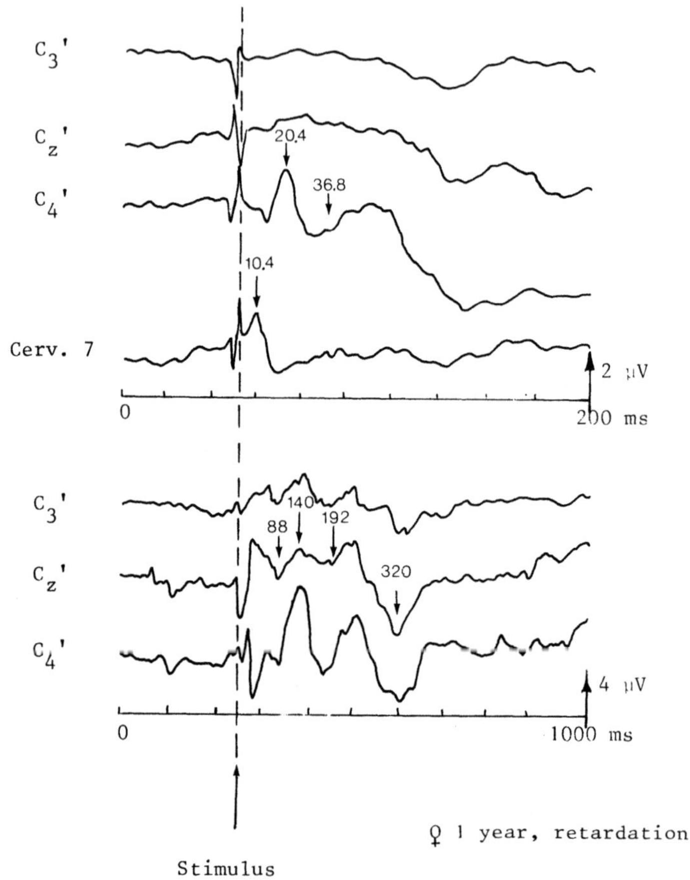

Fig. 5.19

<u>Method</u>

- right index finger
- at random, ¼Hz
- 8 mA, 0.1 ms
- filters 1-250 Hz
- reference at A2.

Cervical	:	N_{15}	=	10.4 ms
Cortical	:	N_{20}	=	20.4 ms
		P_{45}	=	36.8 ms
		P_{100}	=	88 ms
		N_{140}	=	140 ms
		P_{200}	=	192/320 ms

<u>Conclusion</u>: normal SSEP

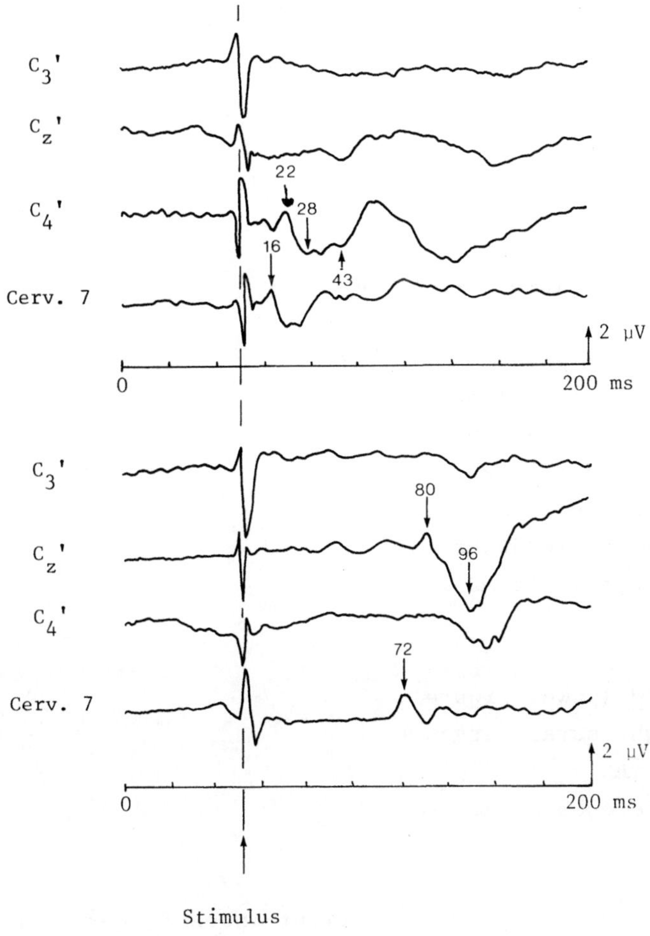

♂ 18 years, myelitis transversa Th4

Fig. 5.20

Fig. 5.20 (see former page)

<u>Method</u> (upper figure):
- right index finger
- at random ¼ Hz
- 11 mA, 0.1 ms
- filters 1-250 Hz
- reference at A2

<u>Method</u> (lower figure):
- right sural nerve
- 13 mA
- reference at A2

	right index finger	right sural nerve
Cervical : N_{15}	16 ms	72 ms
Cortical : N_{20}	22 ms	80 ms
P_{45}	28/34 ms	96 ms

<u>Conclusion</u>: conduction disturbance below the cervical level.

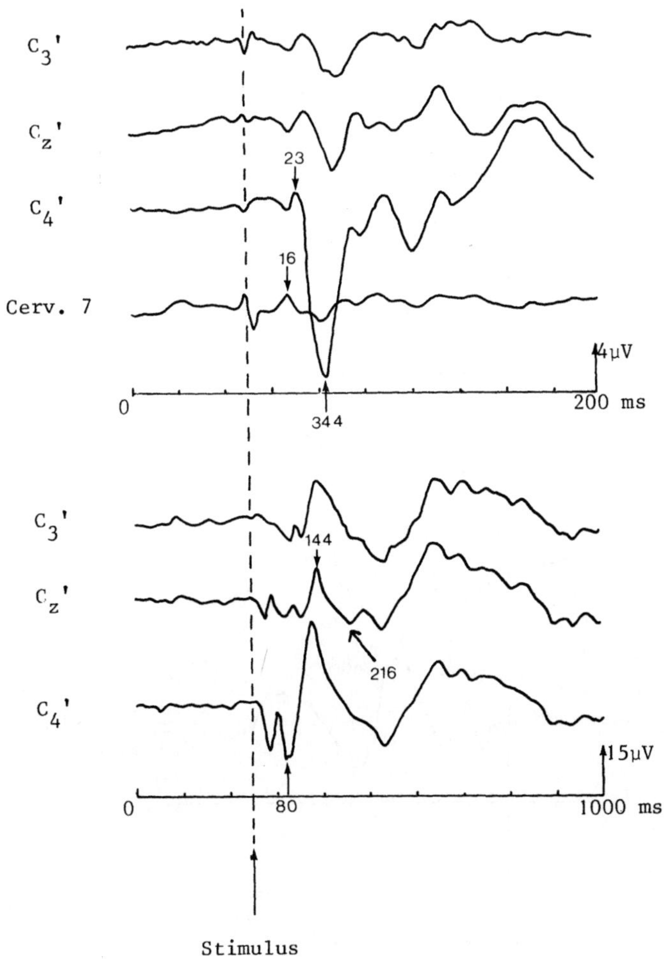

♂ 17 years, hypoparathyreoidia

Fig. 5.21

<u>Method</u>
- right index finger
- at random ¼ Hz
- 9 mA, 0.1 ms
- filters 1-250 Hz
- reference at A2

Cervical:	N_{15}	= 16	ms
Cortical:	N_{20}	= 23	ms
	P_{45}	= 34.4	ms
	P_{100}	= 80	ms
	N_{140}	= 144	ms
	P_{200}	= 216	ms

<u>Conclusion</u>: Abnormal P_{45}, indicating toward local irritation on C_4'. (EEG: spike focus on C_4').

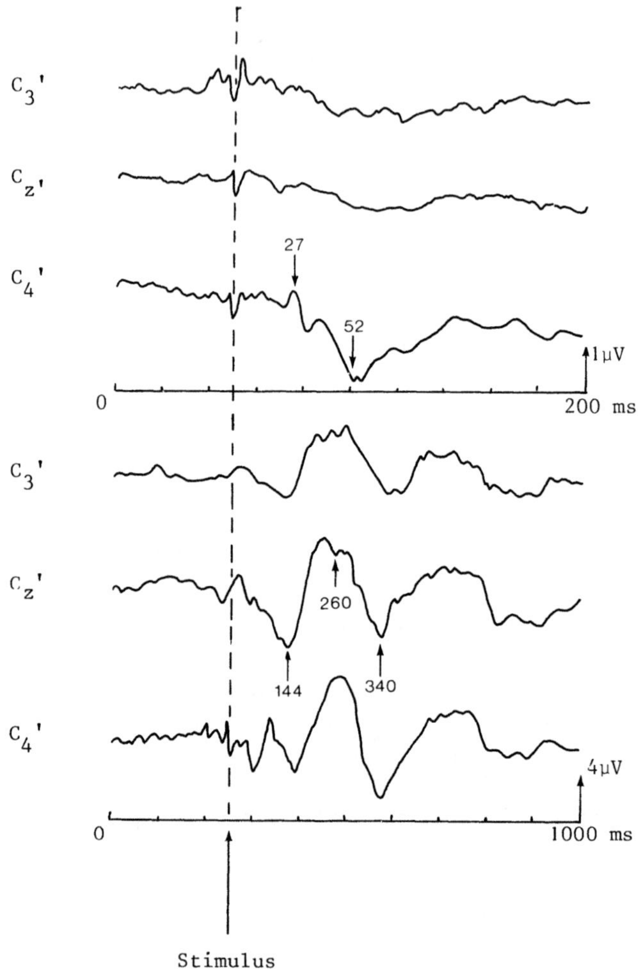

♂ 30 years, Sudanophilic Leucodystrophia

Fig. 5.22

<u>Method</u>

- right index finger
- at random ¼ Hz
- 9 mA, 0.1 ms
- filters 1-250 Hz
- reference at A2

Cortical	N_{20}	=	27 ms
	P_{45}	=	52 ms
	P_{100}	=	144 ms
	N_{140}	=	260 ms
	P_{200}	=	240 ms

<u>Conclusion:</u> white matter disease and probably also "gray matter" (or <u>diffuse</u> cortical) **dysfunction**.

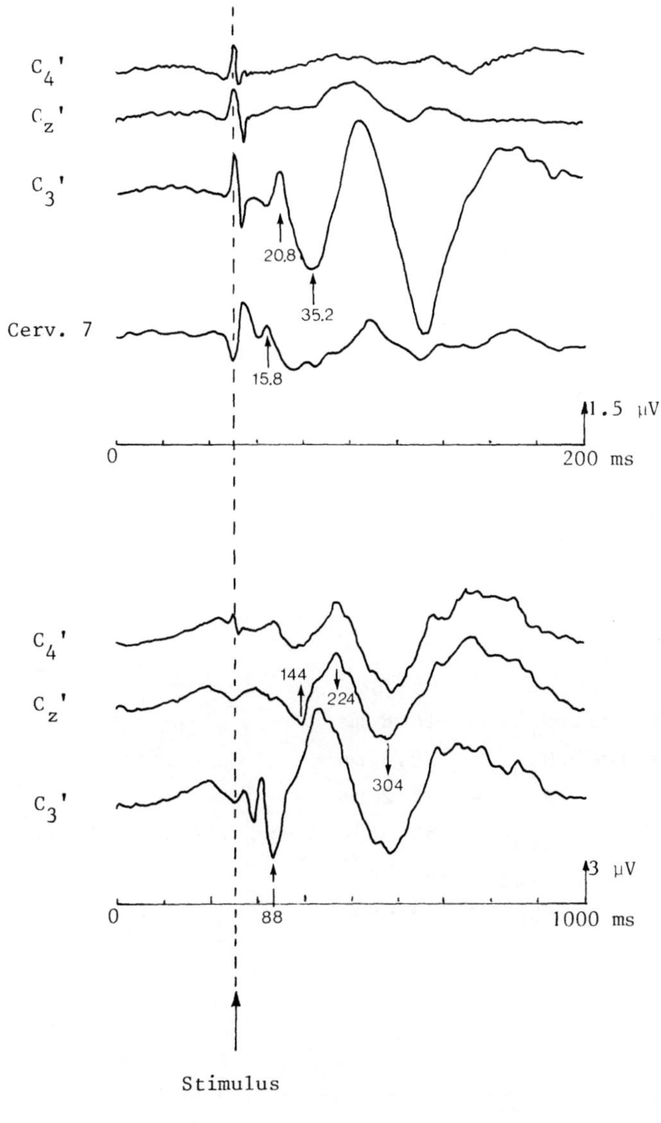

Fig. 5.23 ♀ 7 years, "Minimal Brain Damage"

Fig. 5.23 (see former page)

Method

- right index finger
- at random ¼ Hz
- filters 1-250 Hz
- 14 mA, 0.1 ms
- reference at A2

	Cervical: N_{15}	= 15.8 ms
	Cortical: N_{20}	= 20.8 ms
	P_{45}	= 35.2 ms
Note:	P_{100}	= 88 at C_3' and 144 at C_z' and C_4'
	N_{140}	= 224 ms
	P_{200}	= 304 ms

Conclusion: indication for "gray matter" dysfunction.

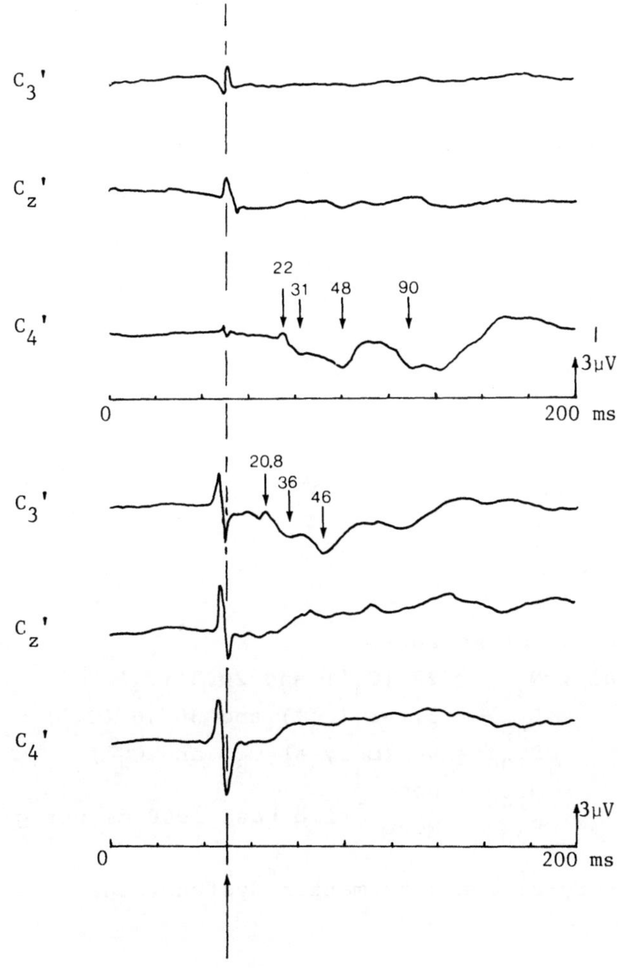

♂ 13 years, SSPE

Fig.5.24

Fig.5.24 (see former page)

<u>Method</u> (upper figure):
- right index finger
- at random ¼ Hz
- 12 mA, 0.1 ms
- filters 1-250 Hz
- reference at A2

<u>Method</u> (lower figure):
- left index finger
- reference at A2

Cervical : not measured

Cortical : N_{20} = 22 (C_4') and 20.8 (C_3')

P_{45} = 31/48 (C_4') and 36/46 (C_3')

P_{100} = 90 (only at C_3' and C_4')

N_{140} = None

P_{200} = None   time base 1000 ms not given here

<u>Conclusion</u>: very severe "gray matter" dysfunction.

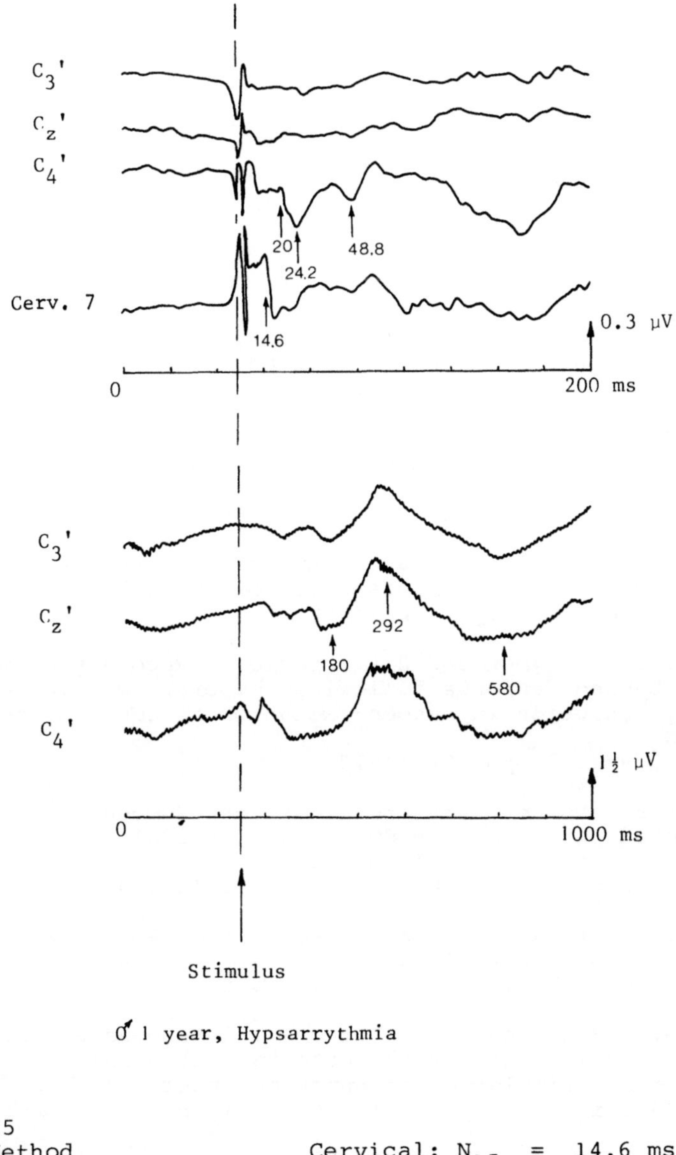

♂ 1 year, Hypsarrythmia

Fig. 5.25

Method	Cervical: N_{15} = 14.6 ms
- right index finger	Cortical: N_{20} = 20 ms
- at random ¼ Hz	P_{45} = 24.2/48.8 ms
- 7 mA, 0.1 ms	P_{100} = 180 ms
- filters 1-250 Hz	N_{140} = 292 ms
- reference at A2	P_{200} = 580 ms

Conclusion: severe "gray matter" dysfunction.

REFERENCES

Anziska B.J. and R.Q. Cracco: Short Latency Somato Sensory Evoked Potentials in brain dead patients. Arch. Neurol., vol. 37, pp. 222-225, (1980).

Anziska B.J. and R.Q. Cracco: Short Latency Somato Sensory Evoked Potentials: studies in patients with focal neurological disease. EEG and Clin. Neurophysiol., vol. 49, pp 227-239, (1980).

Baker J.B., S.J. Larson, A.Sances and P.T. White: Evoked Potentials as an aid to the diagnosis of multiple sclerosis. Neurol, vol. 18, pp. 286 (1968).

Branston N.M., L. Symon and H.A. Crockard: Recovery of the Cortical Evoked Response following temporary middle cerebral artery occlusion in baboon: Relation to local blood flow and PO_2. Stroke, vol. 7, pp. 2, (1976).

Bruggenhout E., P. Ketelaer and H. Carton: Succes and failure of evoked potentials in detecting clinical and subclinical lesions in multiple sclerosis. Clin. Neurol., Neurosurg., 84, pp. 3-14 (1982).

Buchthal F. and A. Rosenfalck: Evoked Action Potentials and conduction velocity in human sensory nerves. Brain Research, vol. 3, no. 1, (1966)

Burke D., N.F. Skuse and A.K. Lethlean: Cutaneous and muscle afferent components of the cerebral potential evoked by electrical stimulation of human peripheral nerves. EEG and Clin. Neurophysiol., vol. 51, pp. 579-588, (1981).

Colon E.J., S.L.H. Notermans, H.M. Vingerhoets, J. Kap and J. de Weerd: Cortical and Cervical Somato Sensory Evoked Responses in demyelinating diseases. European Neurology, vol. 15, pp. 124-130, (1977).

Colon, E.J., E.M.G. Joosten and J.P.C. de Weerd: Somato Sensory Evoked Response in controlled A-alpha sensory fibre disease.
J. of Neurol., vol. 219, pp. 273-278, (1978)

Colon, E.J., W. Renier and F. Gabreëls: Propagation defects in somato sensory evoked responses in children.
H. Lechner and A. Aranibar (Eds.): EEG and Clin. Neurophys., Elsevier, North Holland, pp. 306-313, (1979).

Chiappa, K.H., S.K. Choi and R.R. Young: Short latency somatosensory evoked potentials following median nerve stimulation in patients with neurological lesions.
Prog. Clin. Neurophysiol., vol. 7, pp. 264-281, (1980)
Ed: J.E. Desmedt, Karger Basel, (1980).

Dawson, G.D.: Cerebral responses to electrical stimulation of peripheral nerve in man.
J. of Neurol. and Neurosurg. Psychiatry, vol 10, pp. 134-140, (1947A)

Dawson, G.D.: Investigations on a patient subject to myoclonic seizures after sensory stimulation.
J. Neurol., Neurosurg. Psychiatry, vol. 10 pp. 141-162, (1947B)

Desmedt, J.E.: Somato sensory evoked potentials in man: Maturation, cognitive parameter, and clinical uses in neurological disorders.
Lehmann, D., Callaway, E., eds. Human Evoked Potentials. New York: Plenum Press, pp. 83-103 (1979).

Desmedt, J.E. and E. Brunko: Functional organization of far-field and cortical components of somato sensory evoked potentials in normal adults.
Prog. Clin. Neurophysiol., vol. 7, pp. 27-50, (1980).
Ed: J.E. Desmedt, Karger Basel, (1980).

Desmedt, J.E. and G. Cheron: Prevertebral (oesophageal) recording of subcortical somato sensory EP's in man: the spinal P13 component and the dual nature of the spinal generators.
EEG and Clin. Neurophysiol., vol. 52, pp. 257-275, (1981).

Duffy, F.H., J.L. Burchfiel, C.T. Lombroso: Brain electrical activity mapping (BEAM): a method for extending the clinical utility of EEG and evoked potential data.
Ann. Neurol., vol. 5, pp. 309-321, (1979).

Duffy, F.H., M.B. Denckla, P.H. Bartels et al.: Dyslexia: Automatic diagnosis by computerized classification of brain electrical activity.
Ann. Neurol., vol. 7, pp. 421-428, (1980).

Duffy, F.H.: Brain electrical activity mapping (BEAM): computerized access to complex brain function. Intern. J. Neuroscience, vol. 13, pp. 55-65, (1981).

Eisen, A. and G. Elleker: Sensory nerve stimulation and evoked cerebral potentials. Neurology, vol. 30, pp. 1097-1105, (1980).

Ferris, G.S.: Somato Sensory Evoked Responses in callosal agenesis. Electroenceph. Clin. Neurophys., 35, pp. 442, (1973).

Goff, G.D., Y. Matsumiya, T. Allison and W.R. Goff: The scalp topography of human somato sensory and auditory evoked potentials. EEG and Clin. Neurophysiol., vol. 42 , pp. 57-76, (1977).

Goff, W.R., P.D. Williamson, J.C. van Gilder, T. Allison and T.C. Fisher: Neuronal origins of long latency evoked potentials recorded from the depth and from cortical surface of the brain in man: Clinical uses of cerebral, brainstem and spinal somato sensory evoked potentials. Prog. Clin. Neurophysiol., vol. 7, pp. 126-145, (1980). Ed.: J.E. Desmedt, Karger Basel, (1980).

Goldie, W.D., K.H. Chiappa, R.R. Young and E.B. Brooks: Brainstem Auditory and short latency somato sensory evoked responses in brain-dead. Neurology, vol. 31, pp. 248-255, (1981).

Greenberg, R.P., D.J. Meyer, D.P. Becker and J.D. Miller: Evaluation of brain function in severe human headtrauma with multi-modality evoked potentials. J. of Neurosurg., vol. 47, pp. 150-162, (1977).

Halliday, A.M., and G.S. Wakefield: Cerebral evoked potentials in patients with dissociated sensory loss. J. Neurol. Neurosurg. Psych., vol. 26, pp. 211-219, (1963).

Hrbeck, A., P. Karlberg, D. Kjellmer, T. Olsson and M. Rika: Clinical application of evoked electroencephalographic Responses in newborn infants. I Perinatal Asphyxia. Develop. Med. Child Neurol., vol. 19, pp. 34-44, (1977).

Hubel, T.H. and T.N. Wiesel: Receptive fields, binocular interactions and functional architecture in the cat's visual cortex. J. of Physiol., vol. 160, pp. 106-154, London, (1962).

Hume, A.L., B.R. Cant, N.A. Shaw and J.C. Cowan: Central somatosensory conduction time from 10 to 79 years. EEG and Clin. Neurophysiol. 54, pp. 49-54, (1982).

Jones, S.J.: An interference approach to the study of somato sensory evoked potentials in man. EEG and Clin. Neurophysiol., vol. 52, pp. 517-530, (1981).

Kazis, A., D. Xafenias, J. Papanastasion, A. Doacoyiannis and G. Forogeon: Somato sensory evoked potentials in unilateral cerebral lesions.
Electromyogr. Clin. Neurophysiol., 22, pp. 163-182, (1982).

Krumholz, A., H.D. Weiss, P.J. Goldstein and K.C. Harris: Evoked Responses in Vitamin B12 Deficiency.
Ann. Neurol., vol. 9, pp. 407-409, (1981).

Laget, P., R. Salbreux, J. Ranisbault, A.M. d'Allest and J. Marianic: Relationship between changes in somesthetic evoked responses and electroencephalographic findings in the child with hemiplegia.
Develop. Med. Child Neurol., vol. 18, pp. 620-631, (1976).

Levy, R. and J. Muskin: The somato sensory evoked response in patients with hysterical anaesthesia.
J. Psychosom. Res., vol. 17 (2) pp. 81-84, (1973).

Lesser, R.P., H. Lueders, J. Hahn and G. Klein: Early somatosensory potentials evoked by median nerve stimulation: Intraoperative monitoring.
Neurology (Ny), 31, pp. 519-523, (1981).

Lüders, H.: The effect of aging on the wave form of the somato sensory cortical evoked potential.
EEG and Clin. Neurophysiol., vol. 29 (5), pp. 450-460, (1970).

Ludin, H.P.: Electromyography in practice.
Georg Thieme, Stutgart- New York, (1980).

Markant, O.N., B.Garg, W. DeMeyer, C.Warren and R.M. Worth: Brainstem auditory, visual and somatosensory evoked potentials in leucodystrophies.
EEG and Clin. Neurophysiol., vol. 54, pp. 39-48, (1982).

Matthews, W.B., M. Esiri: Multiple sclerosis plague related to abnormal somato sensory evoked potentials.
J. Neurol. Neurosurg. Psych., vol. 42, pp. 940-942, (1979).

Matthews, W.B., D.G. Small: Serial recording of visual and somato sensory evoked potentials in multiple sclerosis.
J. Neurol. Sci., vol. 40, pp. 11-21, (1979).

Moldofsky, H. and R.S. England: Fascilitation of somato sensory average evoked potentials in hysterical anaesthesia and pain.
Arch. Gen. Psychiatry, vol. 32, pp. 193-197, (1975).

Nakanishi, T., Y. Shimada and Y. Toyokura: Somato sensory evoked responses to mechanical stimulation in normal subjects and in patients with neurological disorders.
J. Neurol., Sciences, vol. 21, pp. 289-298, (1974).

Namerow N.S., R.J.Sclabassi and F.E. Nelson: Somato sensory responses to stimulus trains: Normative data.
EEG and Clin. Neurophysiol., vol. 37, pp. 11-21, (1968).

Noël P. and J.E. Desmedt: Somato sensory cerebral evoked potentials after vascular lesions of the Brain-stem and Diencephalon.
Brain, vol. 98, pp. 113-128, (1975).

Noël P. and J.E. Desmedt: Cerebral and far-field somato-sensory evoked potentials in neurological disorders involving the cervical spinal cord, brainstem, thalamus and cortex. Clin. uses of cerebral, brainstem and spinal somato sensory evoked potentials.
Neurophysiol., vol 7, pp. 205-230, (1980).
Ed.: J.E. Desmedt, Karger Basel, (1980).

Oepen G., M. Doers and U. Thoden: VEP and SSEP in Huntington's Chorea.
EEG and Clin. Neurophysiol., vol. 51, pp. 666-670, (1981)

Riffel, B. and M.Stöhr: Spinale und Subkortikale somatosensorisch evozierte Potentiale nach Stimulation des N. Tibialis.
Arch. Psychiatr. Nervenkr., 232, pp. 251-263, (1982).

Rossini P.M., R.Q. Cracco, J.B. Cracco and W.J. House: Short latency somato sensory evoked potentials to peroneal nerve stimulation: Scalp topography and the effect of different frequency filters.
EEG and Clin. Neurophysiol., vol. 52, pp. 540-552, (1981).

Rushton D.M., J.C. Rothwell and M.D. Craggs: Gating of somato sensory evoked Potentials during different kinds of movement in man.
Brain, vol. 104, pp. 465-491, (1981).

Seppä-äinen A.M., K.Savolainen and T. Kovala: Changes induced by xylene and alcohol in human evoked potentials.
EEG and Clin. Neurophysiol., vol. 51, pp. 148-155, (1981).

Shagass C. and J. Straumanis: Drugs and Human sensory Evoked Potentials.
Lipton M.A., A. Dimascio and K.F. Kielam (Eds): Psychopharmacology, a generation of progress, pp. 699-709, Raven Press, N.Y., (1978).

Shagass C., R.A. Roemer, J.J. Straumanis and M.Amadeo: Topography of sensory evoked potentials in depressive disorders
Biol. Psychiat., 15, 2, pp. 183-207, (1980).

Shibasaki H., Y. Yamashika and Y. Kuroiwa: Electroencephalographic studies of myoclonus.
Brain, vol. 101, pp. 447-460, (1978).

Stöhr, M., U. Buettner, B. Riffel and E. Koletzki:Spinal somatosensory Evoked Potentials in cervical cord lesions. EEG and Clin. Neurophysiol., 54, pp. 257-265, (1982).

Suzuki A., Z. Ito, K.Yoshioka and M. Nara: Topographical Display of evoked potentials; its application to cerebro vascular diseases.
EEG and Clin. Neurophysiol., vol. 52, pp. S42, (1981).

Tackmann W. and D. Kuhlandahl: Evoked Potentials in neuronal ceroid lipofuchsinosis.
Europ. Neurol., vol. 18, pp. 234-242, (1979).

Tamaki M., T. Nakanishi and N. Kudo: An analysis of components of the somato sensory evoked potential by Kaimic Acid in the cat.
EEG and Clin. Neurophysiol., vol. 52, no.3, pp. S82 (1981)

Tamura K., H. Lüders and Y. Kuroiwa: Further observations on the effects of aging on the wave form of the somato sensory cortical evoked potential.
EEG and Clin. Neurophysiol., vol. 33, pp. 325-327, (1972)

de la Torre J.C., J.L. Trimble, R.T. Beard, K. Hanlon and J.W. Surgeon: Somato sensory evoked potentials for the prognosis of coma in humans.
Exp. Neurol., vol. 60, pp. 304-317, (1978).

Trojaborg W., and E. Peterson: Visual and somato sensory evoked cortical potentials in multiple sclerosis.
J. Neurol. Neurosurg. Psych., vol. 42, pp. 323-330, (1979).

Tsumoto T., N. Hirose, S. Nonaka and M.Takanishi: Analysis of somato sensory evoked potentials to lateral popliteal nerve stimulation in man.
EEG and Clin. Neurophysiol., vol. 33, pp. 379-388, (1972).

Yamada T., E.Shivapous, T.T. Wilkinson and J. Kimura: Short and long latency somato sensory evoked potentials in multiple sclerosis.
Arch. Neurol., vol. 39, pp. 88-94, (1982).

de Weerd A.W. : Non-invasief onderzoek bij multiple sclerose patienten; een studie met computertomografie en evoked response onderzoek.
Ph. D. thesis, J.H. Pasmans, Den Haag, the Netherlands. (1981).

de Weerd J.P.C.M.: Estimation of Evoked Potentials; a study of a posteriori "Wiener" Filtering and its Time-varying Generalization.
Ph.D. thesis, Krips repro Meppel, the Netherlands, (1981).

SUBJECT INDEX